HANDBOOK OF
SIMPLIFIED
ELECTRICAL WIRING DESIGN

HANDBOOK OF
SIMPLIFIED
ELECTRICAL WIRING DESIGN

John D. Lenk

Prentice-Hall, Inc., *Englewood Cliffs*, *New Jersey*

Library of Congress Cataloging in Publication Data

LENK, JOHN D
    Handbook of simplified electrical wiring design.

    1. Electric wiring—Handbooks, manuals, etc.
    I. Title.
    TK3271.L45   621.319'24'0202   74-9998
    ISBN 0-13-381723-7

10   9   8   7   6   5   4   3   2

Printed in the United States of America

Prentice-Hall International, Inc., *London*
Prentice-Hall of Australia, Pty. Ltd., *Sydney*
Prentice-Hall of Canada, Ltd., *Toronto*
Prentice-Hall of India Private Limited, *New Delhi*
Prentice-Hall of Japan, Inc., *Tokyo*

To
Irene, Karen,
Mark, Lambie, and Mandy

Contents

Preface

The purpose of this book is to tell a journeyman electrician or working contractor how to solve specific wiring problems. The book assumes that the reader is experienced in making installations planned by others, where all the paperwork problems have been solved, and that the reader now wants to do his own planning and problem solving.

The information in this book tells the experienced working man how to do some of the work of a contractor. The book touches on estimating but avoids the legal and business end of electrical contracting. Instead, the book concentrates on practical design and planning of electrical wiring.

The book assumes that the reader is an experienced electrician, so very few mechanical details are included (only where necessary to solve wiring design problems). The book assumes also that the reader is familiar with basic electrical equations, Ohm's law, magnetism, and the like. No basic electrical theory is repeated. However, a summary of basic electrical equations and how they are applied to practical design is included in the final chapter.

The book investigates capabilities and limitations of various wiring system alternatives so that the reader can make logical choices to meet specific design requirements. The book also provides shortcuts and rules of thumb for those instances where a fast, approximate, design answer will suffice. For example, the chapters on lighting and heating describe the standard methods for calculation of lighting and heating requirements. This includes all necessary tables, together with worked-out examples. The discussion of standard methods is followed by shortcut methods, which can be applied to the great majority of cases.

Chapter 1 discusses the basics of electrical power distribution (generating station to service entrance, and service entrance to interior electrical outlets and loads). This chapter is included as a starting point for practical design, a review for the experienced reader, and as an introduction for the student reader.

Chapter 2 covers raceways and conductors. All necessary tables and equations, plus worked-out examples, are included to calculate raceway and conductor sizes, material, and types.

Chapter 3 discusses the distribution of electrical power from the service entrance to electrical outlets and loads. Included are tables, equations, and examples to help in finding correct feeder, branch circuit, and overcurrent protection sizes.

Chapter 4 covers grounding systems in electrical wiring and describes the how and why of grounding from a design standpoint, with a discussion of NEC grounding requirements. The chapter also discusses ground fault circuit interrupters and associated wiring problems.

Chapter 5 describes how transformers can be used to provide the different voltages required in modern electrical wiring systems and how they are used in low-voltage control systems.

Chapter 6 discusses the design of electrical lighting from the standpoint of providing a given amount of uniform light for a given area. Detailed calculations, both standard long form and shortcuts, are given to determine how many lamps are required and how the lamps should be spaced.

Chapter 7 provides the same type of information as Chapter 6 except that the data apply to electrical heating. Heating requirements and how they are met are discussed from the standpoint of design planning.

Chapter 8 discusses the fundamentals of electrical motor control, including how to select starters and controllers of the proper rating for a given motor. Also included are data on understanding motor control wiring diagrams.

Chapter 9 is a review, or brief summary, of basic electrical data most needed for the planning and design of electrical wiring systems. Basic electrical theory is not included, nor is mathematical analysis. Mathematics is used only where absolutely necessary, and then in the simplest form.

Since the book does not require advanced math or theoretical study, it is ideal for the working electrician or contractor who needs a ready reference source for basic equations, calculation techniques, and related information. The book is suited also for schools that teach electrical power distribution analysis with a strong emphasis on practical, simplified design.

The author has received much information from various organizations and experts in the field. He wishes to thank them all for their help and to express special appreciation to the Lamp Division, Industrial Heating Division, Electric Comfort Heating Department, and Wiring Device Division of General Electric Company; the Industrial Controls Group and the Distribution Equipment Division of the Square D Company; the National Joint Apprenticeship and Training Committee for the Electrical Industry; the Department of Water and Power for the City of Los Angeles; and the Illuminating Engineering Society.

Los Angeles, California   John D. Lenk

Chapter 1

Basics of
Electrical Power
Distribution

In this chapter we shall discuss the basics of electrical power distribution from the generating station to the service entrance of the individual customer, and from the service entrance to the electrical outlets. We shall also discuss the equipment involved with electrical wiring (from a simplified, practical standpoint). The systems and equipment that we treat are typical of those in most general use throughout the industry.

We shall not discuss basic electrical theory in any detail, although a summary of electrical theory in Chapter 9 includes Ohm's law, the calculation of power in single- and three-phase circuits, the power factor, power factor correction, and related subjects necessary for wiring design.

No book on electrical wiring is complete without reference to the National Electrical Code (NEC). Although frequent reference to the Code is made throughout the book, we made no attempt to duplicate the NEC nor even to cover every paragraph in full detail, as copies of the Code are available from most utility companies. Where practical, Code references are made in italic type to assist the reader in becoming quickly familiar with the Code's provisions as they apply to simplified design problems.

1-1 NATIONAL ELECTRICAL CODE

To assure that electrical equipment is standardized and properly installed, the National Fire Protection Association (NFPA) has developed a set of minimum standards for electrical installations in homes, stores, industrial plants, farms, and so on. The NFPA also determines manufacturing standards for many electrical items. These manufacturing standards, which apply primarily to size, assure that lamps of a given type will fit all sockets of the corresponding type, plugs will fit receptacles, and the like. The NFPA standards are not to be confused with those of the Underwriters' Laboratories (UL). The UL is a nationally accepted organization that tests all types of wiring materials and devices to make certain that they meet minimum standards for *safety* and *quality*.

The NFPA standards concerning installation are incorporated into a book called the *National Electrical Code* (NEC). Each section of the NEC is supervised by a panel of experts in various areas. These experts check the code against the latest developments in the electrical industry, revising it as required to reflect changes in modern electrical wiring and equipment. Typically, the NEC is revised every three years.

Although the NEC is a nationally accepted document, all its provisions are subject to local interpretation. Each state, county, city, and township may have a code or regulations for installation of electrical equipment and wiring. In some cases the local authority will accept part of the NEC without change and then publish regulations that supersede other parts of the NEC. A classic example is the use of conduit that houses electrical wiring (see Chapter 2). Many local authorities require either rigid conduit or armored cable for all wiring (both new work and additions). Regulations in other areas permit nonmetallic (typically plastic) covering for cable used in additions, remodeling, and the like.

Another problem is that local wiring inspectors invariably differ in their interpretation of the provisions of the NEC. For these reasons, the NEC provisions should be considered to be *minimum requirements* for any electrical wiring installation. Also, the NEC's provisions apply in the absence of the requirements of a local authority.

The NEC should not be considered to be an instruction manual. However, many NEC standards form the basis for the solution of problems related to the design of electrical wiring. For example, the NEC permits a maximum voltage drop of 5 per cent (of the source voltage) from the service entrance (point where electrical power enters the house or building) to the most distant electrical outlet. This voltage drop limit of 5 per cent sets the size of electrical wiring in each circuit throughout the system. (Voltage drop and wire size are discussed fully in Chapter 3.)

To sum up, anyone involved in the design of electrical wiring should have a copy of the NEC, plus any local codes or regulations. If there is doubt as to which local agency governs the installation of electrical systems, contact the local utility company. In fact, the local utility company should always be contacted in regard to electrical wiring design problems, since each utility company has its own requirements for electrical service (size and type of service entrance equipment, length of service cable, etc.).

1-2 ELECTRICAL POWER DISTRIBUTION TO THE CUSTOMER

The distribution of electrical power from the generating station to the individual customer is shown in Fig. 1-1. Note that this system is "typical." That is, the electrical power is alternating current at 60 hertz (Hz) and is made through three-phase transmission lines. Of course, the distribution of electrical power is not the same for all areas of the country. In addition, the distribution system shown in the figure is the responsibility of the utility company, not of the contractor or electrician. However, the distribution of electrical power to the service entrance is the starting point for the solution of any electrical wiring design problem.

As shown in Fig. 1-1, the generating station (hydroelectric, steam, etc.) produces voltages in the range 12 to 24 kilovolts (kV). This voltage is stepped up (by transformers, as described in Chapter 5) to transmission voltages in the range 60 to 110 kV. The reason for such high transmission voltages is because the voltage drops as the electrical power passes along the transmission cables from generating station to substation. As a guideline, the transmission voltage must be approximately 1000 V for each mile of transmission. Thus a substation 100 miles from the generating station requires transmission voltages of 100 kV or higher.

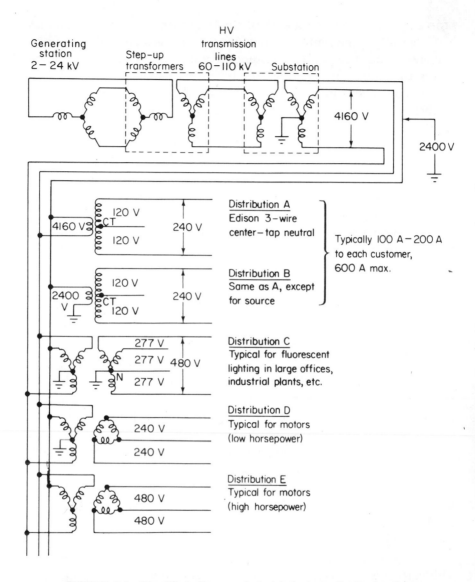

FIGURE 1-1   Simplified diagram of electrical power distribution from generating station to individual customer.

Power is distributed from the substation to individual customers (or groups of customers) through three-phase transmission lines. Typically, the voltage across each phase is 2400 V. Since the transformers at the substation output are connected in three-phase wye configuration, the voltage from phase to phase is 4160 V. (Readers not familiar with three-phase electrical power distribution should refer to Chapters 5 and 9.) The 2400/4160 V power is stepped down by transformers to voltages suitable for each customer's needs. Typical distribution systems are described in the following sections.

1-2.1 120/240-V single-phase distribution

Most lighting and appliances in homes and small buildings require 120 or 240 V. Generally, 240 V is used only for large appliances. You should note here that in practical applications, most appliances will operate on any voltage between 110 and 120 V (or 220 and 240 V). The power supplied by most utility companies is at voltages within these ranges. Often, such power is referred to as 110, 115, or 120 V, depending upon the locality. Multiples are referred to as 220, 230, and 240 V. In this book, we will standardize on 115 or 120 V (and 230 or 240 V). However, the terms 110 and 220 V still exist and are in common use.

The 120/240 V system shown as distribution A in Fig. 1-1 is supplied by a three-wire system. The transformer primary is fed by 4160 V, taken across two lines of the three-phase transmission. This system, sometimes referred to as the *Edison three-wire system*, is typical for at least 90 per cent of homes and small buildings in the United States. (The theory of the three-wire system is discussed in Chapter 9; practical connections and applications are discussed throughout other chapters.)

For the purposes of this chapter it is sufficient to say that there is 120 V from the center tap to either end of the transformer secondary, and 240 V between the ends of the secondary. The three wires (or service conductors, often called the *service drop*) are brought into the customer's service entrance. However, not all customers use all three wires. When there are no major electrical appliances (electric range, clothes dryer, water heater, etc.), 240 V is not needed, so only two wires are used beyond the service entrance.

The center-tap wire (known as the *neutral* or *ground wire*) is always used. The center-tap wire forms a complete circuit with either of the end taps to provide 120 V. It is estimated that about 70 per cent of the homes and apartments in the United States use only two wires (center tap and one end) of the available three-wire system, since only 120 V is needed.

The 120/240-V system shown as distribution B in Fig. 1-1 is

identical to distribution A in that the customer has been provided with a 120/240-V three-wire Edison system at his service entrance, but in this case the utility company has tapped across one phase of the three-phase transmission line. Thus the transformer primary is fed by 2400 V rather than by 4160 V, which makes it possible to use smaller transformers but provides less power. The system of distribution B is generally used when the distance between transformer and service entrance is less than 600 feet (ft) and no more than six 100-ampere (A) service entrances are served. However, this is the concern of the utility company and not of the designer of the customer's electrical wiring.

1-2.2 277/480-V three-phase distribution

The 277/480-V system shown as distribution C in Fig. 1-1 is supplied by a four-wire system. The transformer primary (connected in wye configuration) is fed by all three phases of the 2400/4160-V transmission. The transformer secondary (also connected in wye) delivers 277 V across each phase and 480 V between phases. The center tap of all three-phase windings is grounded and requires a fourth neutral (or ground) conductor to the service entrance.

The 277/480-V distribution (also referred to as 265/460-V power) is used primarily for the lighting of large industrial plants and office buildings. Many fluorescent lighting systems are designed for use with 277/480-V power.

The 277/480-V distribution has the disadvantage of requiring four conductors rather than three. However, the higher voltage (277 V versus 120 V) permits longer runs for conductors, or lower currents, or a combination. (The advantages and limitations of 277/480-V power are discussed at length throughout several chapters.) The main disadvantage (in addition to requiring four conductors) is that most customers also require 120 V for conventional (nonfluorescent) lighting and other applications. Utility companies generally prefer (and provide) reduced rates when they supply only one type of distribution. This means that the customer must supply transformers (at his own expense) to convert the 277/480-V three-phase to a 120-V single-phase distribution (or possibly to a 120/240-V three-wire Edison system).

1-2.3 240-V and 480-V three-phase distributions

The 240-V and 480-V systems shown as distributions D and E, respectively, in Fig. 1-1, are supplied by three-wire systems. The transformer primary (connected in wye) is fed by all three phases of the 2400/4160-V transmission. The transformer secondary (connected in delta) delivers 240 V (or 480 V) across each phase. No ground or neutral is required; thus only three service conductors are needed.

The 240-V and 480-V three-phase systems are used primarily for electric motors or similar applications. Single-phase electric motors are impractical for any application except the fractional-horsepower type found in home use (vacuum cleaners, mixers, refrigerators, etc.). Motors used in industry, large buildings, and so on, are operated most efficiently when three-phase is used at voltages of 240 V and 480 V (in some cases 600 V).

When the only supply from the utility company is 240 or 480 V, the customer must provide a transformer to convert the 240-V or 480-V three-phase to 120-V single-phase for conventional lighting and other applications.

1-2.4 High-voltage distribution

In addition to the distributions shown in Fig. 1-1, most utility companies also supply higher voltages for special applications. These voltages are typically on the order of 2400, 4800, 4160, 12,470, and 34,500 V. However, high voltages are the exception rather than the rule, so they will not be considered in this book.

1-3 SERVICE ENTRANCE

Although there is an infinite variety of service entrances and equipment (for each type of customer and at each locality), all service entrances have certain elements in common. Any service entrance must provide the necessary conductors, watthour meters, main disconnect means, overcurrent devices (fuses or circuit breakers), and (as discussed in Chapter 4) a system ground.

Each utility company has its own requirements as to service

entrance equipment.   Generally, the utility company will supply on request full details of these requirements.   Design and construction of service entrance equipment and systems will not be discussed in full, but in the following sections we shall summarize service entrance requirements, particularly as related to design and construction.

1-3.1 Types of service entrances

There are two basic types of service entrances: overhead and underground.   Although the overhead entrance is still used in the majority of installations, the underground service entrance is being used more and more frequently.   As noted above, the requirements for both types are set by the particular utility company.   Although underground service entrances are sometimes installed by the utility company, in the case of an overhead installation, the utility company will generally secure the service drop but leave the installation of the service entrance equipment to the customer.

1-3.2 Service drop

The service drop consists of those conductors that connect the utility company's transmission lines to the customer's service entrance equipment.   Figure 1-2 shows some typical requirements for a service drop.   Figure 1-3 shows the attachment risers for a typical conduit service drop.   Keep in mind that these requirements are for one particular utility company and that, in addition, they apply only to overhead service drops for residential buildings.   There are corresponding requirements for other types of service drops.

Although the designers of electrical wiring installations need not design the service entrance equipment, they must follow the requirements, and in so doing, many of the customer's wiring design requirements are set.   For example, to meet the requirements of a service drop location (Fig. 1-2), it may be necessary to install the service entrance at only one location in a building.   This means that all electrical wiring must be designed from that location.

1-3.3 Service conductors

Figure 1-4 shows the basic electrical wiring requirements at a typical service entrance.   The system shown is for a single-family resi-

THESE MINIMUM CLEARANCES APPLY TO THE LOWEST POINT OF SERVICE DROP SAG

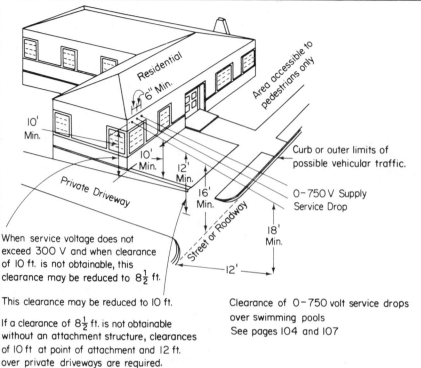

Residential

6" Min.

10'
Min.

10'
Min.

12'
Min.

16'
Min.

18'
Min.

12'

Area accessible to pedestrians only

Private Driveway

Street or Roadway

Curb or outer limits of possible vehicular traffic.

0-750 V Supply
Service Drop

When service voltage does not exceed 300 V and when clearance of 10 ft. is not obtainable, this clearance may be reduced to $8\frac{1}{2}$ ft.

This clearance may be reduced to 10 ft.

If a clearance of $8\frac{1}{2}$ ft. is not obtainable without an attachment structure, clearances of 10 ft at point of attachment and 12 ft. over private driveways are required.

Clearance of 0-750 volt service drops over swimming pools
See pages 104 and 107

MINIMUM ALLOWABLE VERTICAL CLEARANCE OVER BUILDINGS AND STRUCTURES
Service drops of 300-750 volts shall be maintained at a vertical clearance of not less than 8 feet over all buildings and structures.

MINIMUM CLEARANCE FOR 0-300 V SERVICE DROPS ABOVE RESIDENTIAL BUILDINGS				
TYPE OF ROOF		BUILDING SERVED	OTHER BUILDINGS ON PREMISES SERVED	BUILDINGS ON OTHER PREMISES
Approx. 37°	Metal Roof Less than 37°	8 ft.	8 ft.	8 ft.
Rise 9" Run=12"	Metal Roof 37° or more	2 ft.	2 ft.	8 ft.
	Nonmetallic Roof Less than 37°	No limit specified but greatest prac-	2 ft.	8 ft.
	Nonmetallic Roof 37° or more	ticable clearance should be obtained	2 ft.	2 ft.

FIGURE 1-2   Typical requirements for a service drop (*Courtesy* Department of Water and Power, City of Los Angeles).

No couplings will be permitted between the top of the riser and the lowest point of support

See page 105 for service head location

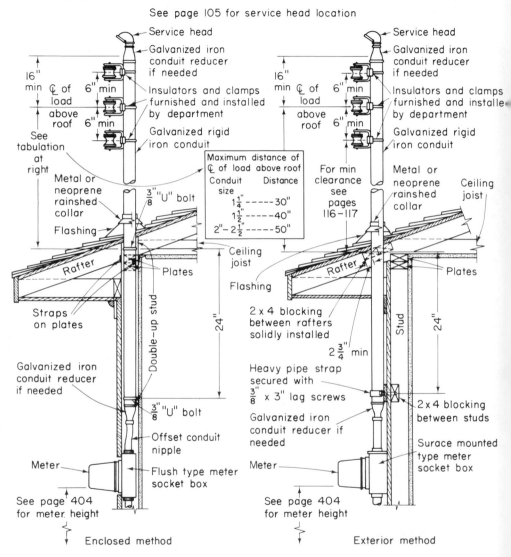

FIGURE 1-3   Attachment risers for a typical conduit service drop
(*Courtesy* Department of Water and Power, City of Los Angeles).

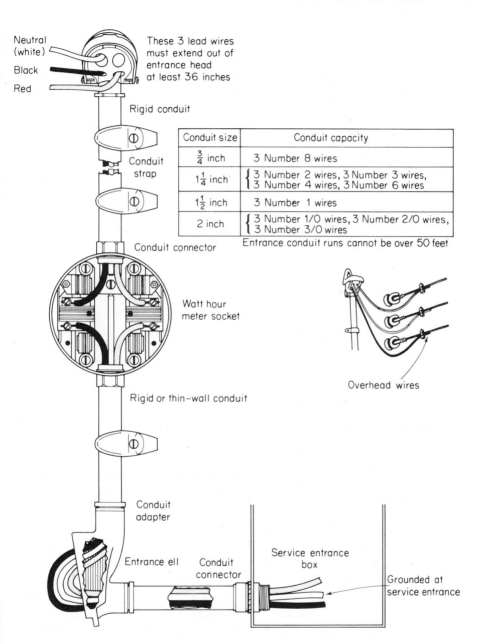

Neutral
(white)

Black

Red

These 3 lead wires
must extend out of
entrance head
at least 36 inches

Rigid conduit

Conduit
strap

Conduit connector

Watt hour
meter socket

Rigid or thin-wall conduit

Conduit
adapter

Entrance ell   Conduit
connector

Service entrance
box

Grounded at
service entrance

Overhead wires

Conduit size	Conduit capacity
$\frac{3}{4}$ inch	3 Number 8 wires
$1\frac{1}{4}$ inch	{ 3 Number 2 wires, 3 Number 3 wires, 3 Number 4 wires, 3 Number 6 wires
$1\frac{1}{2}$ inch	3 Number 1 wires
2 inch	{ 3 Number 1/0 wires, 3 Number 2/0 wires, 3 Number 3/0 wires

Entrance conduit runs cannot be over 50 feet

FIGURE 1-4   Basic electrical wiring requirements at a typical service
entrance (*Courtesy* Department of Water and Power, City of Los Angeles).

dence supplied by the standard three-wire single-phase 120/240-V distribution.

The utility company's overhead wires are connected to the service entrance through the entrance head. In any installation, the entrance head must be higher than the highest or top insulator, and the conductors must extend beyond the entrance head at least 36 inches (in.) and be positioned to form drip loops. These loops prevent water from entering the service.

Typically, the three service conductors are colored as follows: white (for the neutral or ground wire), black, and red (or possibly orange). Typical wire sizes and corresponding conduit sizes are also shown in Fig. 1-4. (Wire and conduit sizes, as well as current capacities, are discussed fully in Chapter 2.)

In this type of service entrance, the conductors are connected to the watthour meter before they reach the service entrance box. This is typical for single-family residences. In multiunit residences and in commercial and industrial installations, the service conductors reach a main disconnect first and are then routed to meters and overcurrent devices.

1-3.4 Service entrance box

Some form of service entrance box is included in all types of service entrances. The service entrance box is known alternatively as the fuse box, the main box, the switch box, the pull box, the circuit breaker box, and the meter box. Some of these names stem from older types of service entrances. In other cases the names apply to different types of service entrances.

Figure 1-5 shows the wiring for a typical service entrance box used in single-family residences. This box has two pull-type disconnects and four fuses. The white (neutral or ground) wire remains connected all times. The black and red wires from the meter are connected to all circuits through the main disconnect. (When the main disconnect is pulled, all power ceases.) The range disconnect provides a means of disconnecting 240 V to the electric range. Each of the four fuses provides protection for the corresponding one of four 120-V circuits.

Figure 1-6 shows a single-family residence service box that uses circuit breakers instead of fuses.

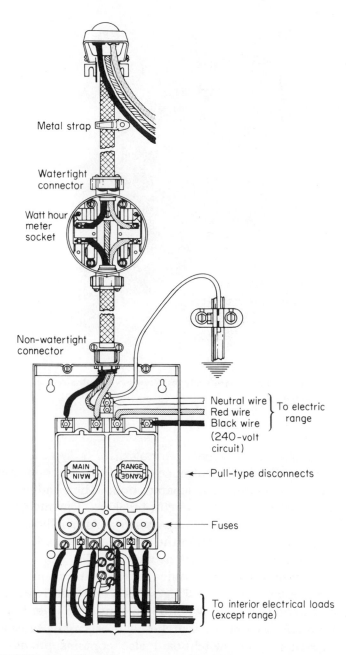

Metal strap

Watertight connector

Watt hour meter socket

Non-watertight connector

Neutral wire ⎫ To electric
Red wire ⎬ range
Black wire ⎭
(240-volt circuit)

MAIN

RANGE

Pull-type disconnects

Fuses

To interior electrical loads (except range)

FIGURE 1-5   Wiring for a typical service entrance box used in a single family residence (*Courtesy* Department of Water and Power, City of Los Angeles).

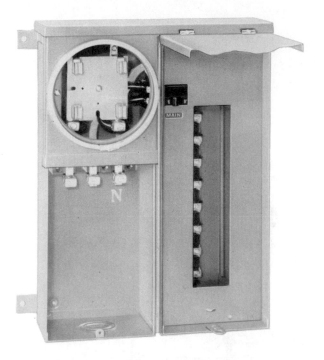

FIGURE 1-6   Typical single-family residence service entrance (*Courtesy* Square D Company).

Figure 1-7 shows the service entrance for a typical multiunit apartment. Here, the service entrance conductors are connected to a main pull-type disconnect in a separate box. The conductors then pass to a number of meters and circuit breakers. One meter and circuit breaker is provided for each apartment. A panelboard (or circuit breaker board) located within the apartment provides additional control and overcurrent protection, as shown in Fig. 1-8.

Figure 1-9 shows a typical service entrance for an industrial installation or similar building. This service entrance is commonly known as a *switchboard*. The switchboard contains a lever-type main disconnect switch, as well as switches and/or circuit breakers for additional circuits. The switchboard also contains provisions for metering (including the watthour meter and current transformers for use with the meter).

FIGURE **1-7**   Typical multi-unit residence service entrance (*Courtesy* Square D Company).

1-3.5 Grounding the service entrance

The subject of grounding is discussed fully in Chapter 4. For the purpose of this chapter it is sufficient to say that the neutral or ground wire of any service entrance must be *grounded at the service entrance.*

1-3.6 Service entrance capacity

Most utility companies do not permit single-phase service entrance installations with capacities in excess of 600 A. When current require-

FIGURE 1-8   Typical panel board used in individual apartments (*Courtesy* Square D Company).

ments are greater than 800 A, the service entrance usually uses *bus bars* rather than wires, as discussed in Chapter 2.

Typically, the service entrance capacity is 200 A or less for single-family homes and small apartment buildings. The following are typical single-family service entrance capacities:

1. *30-A service*, which has been virtually eliminated for any application, with the possible exception of older single-family

FIGURE 1-9   Switchboard serves as service entrance for an industrial installation (*Courtesy* Square D Company).

homes, provides only limited capacity for lighting and for a few of the smaller appliances. This service should never be used except for temporary service or one-room buildings.

2. *60-A service* was standard for many years but is now inadequate for most uses. This service provides the capacity for lighting and for portable appliances, including ranges, low-speed dryers, or hot water heaters, but no additional major appliances can be added at any time.

3. *100-A service* is the *minimum* service required by the NEC for homes up to 3000 ft^2 in floor area. In most localities the minimum service for all new home construction is 100 A. Figure 1-10 shows a typical 100-A service entrance for a single-family home that uses circuit breakers.

4. *150-A service* is preferred for most new home construction.

FIGURE 1-10   Typical 100A service entrance (with circuit breakers) for a single-family residence (*Courtesy* Square D Company).

In homes equipped with an electric range, water heater, high-speed dryer, or central air conditioning, together with lighting and the usual small appliances, 150-A service is minimum.

5. *200-A service* provides the same capacity as 150-A service but will also handle electric house-heating equipment.

The subject of service entrance capacities is discussed further in Chapter 3.

1-3.7 Overcurrent protection

There are two basic types of overcurrent protection in common use, fuses and circuit breakers. These devices protect circuits and wires against overloads and short circuits. A third type of overcurrent device, known as a *ground fault circuit interrupter* (GFCI) or simply *ground fault interrupter*, detects small amounts of current leakage and interrupts the circuit when the leakage reaches a certain level. A complete discussion of the relatively new GFCI is found in Chapter 4. In the present chapter we shall concentrate on conventional fuses and circuit breakers.

Fuses. Fuses are self-destructive overcurrent devices. That is, fuses are destroyed when they perform their function of interrupting the circuit during an overload. Fuses are made of a metal with a low-melting temperature. The amount of fuse metal is calibrated so that it will melt at a specific current rating. Fuses are connected in series with the load so that they will open the circuit when the fuse metal melts.

Fuses have an inverse time characteristic: the higher the overload, the shorter the time before the fuse melts (or "blows"). Some fuses are designed so that there is a time lag before they blow (and interrupt the circuit). This permits overloads of short duration (typically a few seconds) to occur without blowing the fuse.

The most common fuse found in household use is the *screw* or *plug type.* Such fuses are housed in screw shells that fit standard-sized lamp sockets. Screw-type fuses have ratings up to 30 A and are to be used with voltages up to 125 V.

Ferrule- and knife-blade-type fuses are used for higher currents and voltages. Figure 1-11 shows the characteristics of typical fuses. In some power-type switchboards (used in heavy-duty industrial applications) fuses are combined with switches, resulting in a device known as a *fusable switch* (Fig. 1-12).

Circuit breakers. Circuit breakers are reusable overcurrent devices. Circuit breakers interrupt a circuit under overload conditions and can then be reset. Some circuit breakers operate by magnetic means, whereas other units are thermally operated. Most circuit breakers open the circuit within a fraction of a second during overload. There are circuit breakers (usually of the thermal type) that are "tripped" (open the circuit) after a specific time interval.

Circuit breakers are generally available with one, two, or three poles, all mounted on the same assembly. This permits up to three lines or conductors to be interrupted simultaneously. As discussed in Chapter 4, the ground or neutral wire of any service is *never* provided with a circuit breaker (or fuse).

Circuit breakers are available in a wide range of capacities (generally 5 A up to 1000 A). Several manufacturers provide circuit breakers of different capacities that can be inserted into a preassembled

Ferrule			Knife blade		
Amperes	Length	Diameter (inches)	Amperes	Length	Diameter (inches)
30(250 V)	2	9/16	70–100 (250V)	5–7/8	1
30(600 V)	5	13/16	70–100 (600V)	7–7/8	1–1/2
35–60 (250 V)	3	13/16	110–200 (250V)	7–7/8	1–1/2
35–60 (600 V)	5–1/2	1–1/16	110–200 (600V)	9–5/8	1–3/4
			225-400 (250V)	8–5/8	2
			225-400 (600V)	11–5/8	2–1/2
			450-600 (250V)	10–3/8	2–1/2
			450-600 (600V)	13–3/8	3

Screw (plug) type fuses	
Voltage	Amperes
125 V (max)	30 (max)

FIGURE **1-11**   **Characteristics of fuses.**

enclosure. A main or master circuit breaker may be incorporated into the same enclosure. As shown in Fig. 1-10, the enclosures are usually provided with extra space for additional circuit breakers (to be added at a future time as load requirements increase).

1-3.8 Metering

All service entrances include at least one meter to record in kilowatthours the energy consumed by the load. (Kilowatthours and metering are discussed in Chapter 9.) Figure 1-13 shows the basic connections for typical metering. For single-phase 120/240-V service, the meter has two current coils and one voltage coil. For three-phase three-wire service, two voltage coils and two current coils are required. For three-phase four-wire service, three current coils and three voltage coils are needed.

Most kilowatthour meters are plugged into meter sockets. Figure 1-13 also shows typical meter socket jaw arrangements for various services. Note that some of the arrangements are for *totalized meter-*

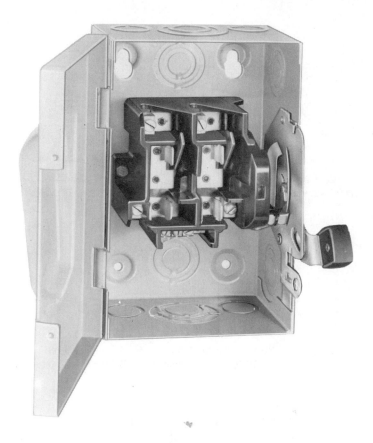

FIGURE 1-12   Fusable switch (*Courtesy* Square D Company).

ing. Some utility companies offer reduced rates when all service in one building can be totalized on one meter. In some cases it is more economical or practical for the customer to receive more than one type of service (120/240-V, 277/480-V, single-phase, three-wire, three-phase, three-wire, etc.) from the utility company. (As discussed in Chapter 5, the alternative to multiple service is for the customer to use a transformer to convert one type of service to other services.) The utility company may require separate meters, or totalized meters, when it provides more than one type of service.

When services are rated over 200 A, *current transformers* are used with the kilowatthour meters to reduce the physical size of the meter. The theory of current transformers is discussed in Chapter 5.

Self contained

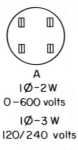

A
1∅-2W
0-600 volts

1∅-3W
120/240 volts

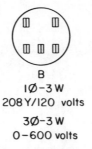

B
1∅-3W
208Y/120 volts

3∅-3W
0-600 volts

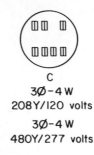

C
3∅-4W
208Y/120 volts

3∅-4W
480Y/277 volts

With instrument transformers

D
1∅-3W

E
3∅-3W
(2) 1∅-3 W (totalized)

F
3∅-4W
(3) 1∅-3 W (totalized)

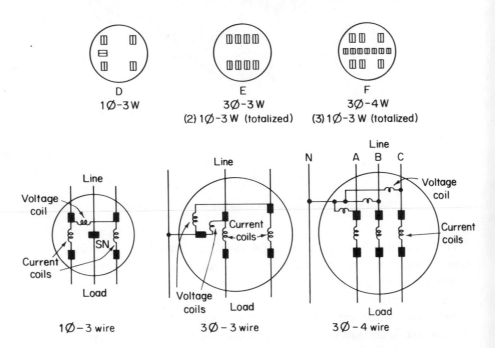

1∅-3 wire

3∅-3 wire

3∅-4 wire

FIGURE 1-13   Typical meter socket jaw arrangements for various services (*Courtesy* Department of Water and Power, City of Los Angeles).

1-3.9 Power factor correction

Utility companies generally require that the power factor for all types of service be at some value not less than about 0.8 to 0.85.

Some utility companies add a surcharge for power factors below 0.8. Power factor correction is usually done by the addition of capacitors located at the service entrance or by means of a synchronous motor. The theory of power factor and power factor correction is discussed in Chapter 9.

1-3.10 Feeders and branch circuits

Electrical power distribution from the service entrance is made by feeder and branch circuits (and possibly panelboards) to the ultimate load or loads. (Feeders and branch circuits are discussed fully in Chapter 3.)

Branch circuit. A branch circuit can be defined as a set of conductors that extend beyond the last overcurrent device in the system. Typically, a branch circuit supplies only a small part of the system.

Feeder. A feeder is a set of conductors that supply a group of branch circuits. In a simple distribution system such as that found in most single-family homes, there are no feeders. The branch circuits are supplied directly from the service entrance, through overcurrent devices (fuses or circuit breakers) located at the service entrance. In more complex systems, such as found in multifamily residences, as well as in commercial and industrial buildings, the branch circuits are supplied through feeders. The feeders receive their power through overcurrent devices at the service entrance and deliver power to the branch circuits through overcurrent devices on *panelboards*. These panelboards are located at various points throughout the building.

Panelboards. Panelboards are assemblies of overcurrent devices contained in a cabinet or panel accessible only from the front. The number of overcurrent devices on a panelboard is limited to 42. These can be either for feeders or branch circuits.

Subfeeders. When many feeders are necessary, a second switchboard (or panelboard, or circuit breaker board) is installed at some point within the building away from the service entrance. This switchboard is supplied with a feeder from the service entrance and supplies the various branch circuit panel boards through *subfeeders*.

1-3.11 Basic branch circuit wiring

Figures 1-14 and 1-15 show the wiring for portions of typical branch circuit panelboards. Figure 1-14 shows the wiring for a 120/

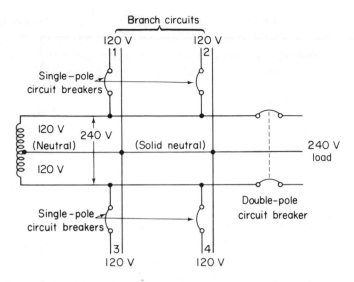

120 V branch circuit loads 1 and 2 = loads 3 and 4 (approx)

FIGURE 1-14   Typical 120-240V, single-phase, 3-wire system.

240-V single-phase three-wire system. There are two basic concerns in design (aside from the current capacities described in Chapter 3).

The system should be balanced. That is, both of the 120-V distributions should have *approximately the same load.* Most utility companies require that the loads be "reasonably balanced" on either side of the neutral conductor. In Fig. 1-14 each 120-V distribution supplies four branch circuits.

The 120-V distribution requires only a single-pole overcurrent device (circuit breaker in this case) for each branch circuit. However, the 240-V distribution requires a double-pole circuit breaker. *Note that the neutral conductor is never provided with an overcurrent device in any distribution.*

Figure 1-15 shows the wiring for a 277/480-V three-phase four-wire system. The basic concern here is that there be a reasonable balance *between each phase of the three-phases.* In Fig. 1-15 each 277-V phase supplies three branch circuits. Again, each 277-V branch circuit requires only a single-pole overcurrent device. However, the three-phase 480-V distribution requires a three-pole overcurrent device.

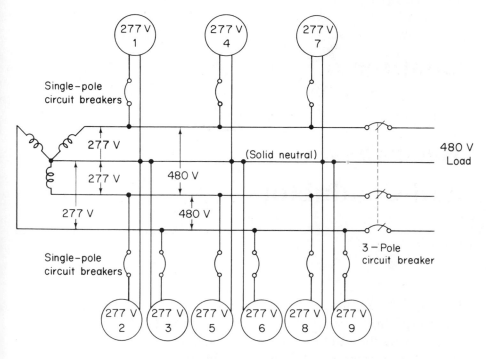

277 V branch circuit loads 1,4 and 7 = 2,5,8 = 3,6,9 (Approx)

FIGURE 1-15   Typical 277-480V, 3-phase, 4-wire system.

Chapter 2

Raceways
and Conductors

In Chapter 1 we discussed how electrical power is transmitted from the generating station to the service entrance of the individual customer by the utility company. In Chapter 3 we shall discuss how the power is transmitted from the individual customer's service entrance to his electrical outlets (for use with lights, heaters, appliances, motors, etc.). In this chapter we shall discuss the *raceways* and *conductors* required for the distribution of power at the customer's location.

It is obvious that such electrical power must be carried by conductors of suitable size and type. Not so obvious, but equally important, however, is the fact that the conductors must be installed in suitable raceways.

2-1 WHY RACEWAYS ARE REQUIRED

Electrical conductors must be protected from heat, moisture and physical abuse. Of course, part of this protection is provided by insulation, but that is not enough. For example, should insulation become even slightly frayed (because of heat or abuse), there is an immediate shock hazard. If there is moisture near the frayed insulation, the moisture could conduct electrical current and cause the wiring to malfunction (or possibly burn).

These possibilities are minimized by installing all conductors in raceways, which can be considered as enclosed channels for wiring. Because of their importance, *the NEC requires that all conductors be enclosed and/or protected by some form of raceway*.

The raceways are mechanically installed as a complete system with all necessary switch boxes, outlets, receptacles, etc. In some cases the conductors must be pulled through the raceways after installation. In other, more modern, electrical wiring systems, the conductors and raceways are manufactured as a single unit (or complete cable).

2-2 TYPES OF RACEWAYS

The choice of raceway type is sometimes set strictly by local electrical codes such that, in effect, there is no choice. In other cases there are several approved raceway types, so that cost is the determining factor. (Note that this cost factor involves both the cost of the raceway material and the cost of labor for installing the raceway.) The following paragraphs summarize the types of raceways available.

2-2.1 Conduit and tubing

Metal tubing of one form or another has long been used for raceways (see Fig. 1-4). There are two types of tubing in common use as raceways: rigid conduit and electrical metallic tubing.

Rigid conduit looks like plumber's pipe and must be handled in about the same way. That is, rigid conduit must be bent to the desired routing (using plumber's pipe-bending tools) and must be threaded at each end. Rigid conduit is attached to other raceway equipment (switch boxes, outlets, etc.) by using bushings and locknuts on the threaded ends. Generally, rigid conduit is available in 10-ft lengths. The main difference between rigid conduit and plumber's pipe is that the interior of the conduit has a heavy enameled finish. This smooth finish permits the conductors to be pulled through the conduit with a minimum of effort.

The main advantage of rigid conduit is its strength. Like plumber's pipe, rigid conduit can be used in areas where there is a chance of physical abuse or where there is extreme moisture. Unlike many other types of raceways, rigid conduit can be buried in concrete when required.

Electrical metallic tubing (EMT) is similar to rigid conduit except that EMT is much thinner and can not be threaded at the ends. EMT (sometimes known as thin wall) is secured to other raceway components by compression rings and/or set screws. The maximum permitted inside diameter of EMT is 4 in. EMT is almost as strong as rigid conduit, and EMT can be buried in concrete. However, EMT is not to be used where there is *continuous* moisture. In addition, special bending tools are required for EMT.

FIGURE 2-1   Construction of typical metal-clad (armored) cable (*Courtesy* Department of Water and Power, City of Los Angeles).

2-2.2 Metal-clad cable

There are two types of *metal-clad* or *armored cables* (see Fig. 2-1) used as raceways: AC and MC. Both types are manufactured as complete cables with the conductors already installed. This eliminates the need to pull the conductors through the raceway. The outer shield of the cables is formed of a spiral steel wrapping. This type of construction makes the raceway, and the conductors, into a flexible cable that can be secured by clamps at various points along its route. The metal-clad cables are attached to switch boxes, outlets, and so on, by means of clamps. Special care must be taken to protect the conductors from the sharp edges of the spiral wrapping at the point where metal-clad cable terminates. Special bushings and fittings are usually required for this purpose.

Metal-clad cable cannot be buried in concrete. One of the reasons for raceways is that conductors can be replaced easily. In metal-clad cable, the conductors and raceway are an inseparable unit, and it is impossible to replace the conductors by pulling them through the armor wrapping. Regular metal-clad cable is not to be used in areas of continuous moisture. However, there is a special metal-clad cable with lead-covered conductors that can be used in high-moisture areas. This metal-clad cable is designated ACL.

The designations AC and MC in metal-clad cable refer to size. MC is used for large loads only, since the smallest conductors available in MC cables are No. 4 AWG (American Wire Gauge) for copper and No. 2 AWG for aluminum. AC is used for smaller loads and is available with conductors as small as No. 14 AWG. Note that AC cable is generally called *BX cable.*

Typically, BX cable is available in two-conductor and three-conductor assemblies. With either type, one conductor must be white (to identify it as the grounded conductor, as discussed in Chapter 4). Generally, the other conductor in a two-conductor cable will be black or red. In three-conductor cables, the other conductors will usually be black and red or orange and blue.

FIGURE 2-2   Construction of typical non-metallic (plastic sheathed) cable (*Courtesy* Department of Water and Power, City of Los Angeles).

2-2.3 Nonmetallic cable (Romex)

Nonmetallic cable (see Fig. 2-2) is generally called *Romex*, although not all such cable may be true Romex. Nonmetallic cable is also manufactured as a complete assembly with the conductors already installed. However, the outer shield is made of plastic or a heavy fabric. Nonmetallic cable is installed in the same way as metal-clad cable but will not take physical abuse without damage. *For this reason nonmetallic cable does not meet the requirements of many local codes.*

For example, nonmetallic cable cannot be used where there is any possibility of physical abuse or in high-moisture areas. Since the outer shield or armor is nonmetallic, it does not conduct current. Thus the armor cannot be used to conduct ground "fault currents," as described in Chapter 4. To overcome this problem, some nonmetallic cable is provided with an extra conductor (usually smaller than the remaining conductors) for grounding purposes.

Typically, nonmetallic cable is available in two-conductor and three-conductor assemblies, as is metal-clad cable. Nonmetallic cable is generally available with conductors in sizes No. 14 to 1 AWG (copper) and No. 12 to 2 AWG (aluminum).

2-2.4 Wireways

Wireways (see Fig. 2-3) are essentially metal channels with removable panels.   In effect, wireways are *special-purpose raceways* designed to meet a number of specific applications.   For example, there are wireways called *surface extensions*, decorative sheet-metal channels designed for routing wires along surfaces (such as walls, floors, overheads, etc.).   Generally, surface extensions are used as extensions of other raceways rather than as full raceway systems designed to wire a complete building.

Wireways are particularly useful in factory and industrial applications.   For example, one overhead wireway consists of a metal duct

FIGURE 2-3   Typical wireway used in industrial applications (*Courtesy* Square D Company).

with bus bars (instead of wires) and openings at various points along the duct. The openings permit switches and outlet to be plugged into the bus bars, as required. Another wireway is used primarily for concrete floors or any underfloor application where there may be considerable moisture.

The main advantage of wireways is that they permit the addition of outlets without extensive rewiring.

2-3 CHOOSING RACEWAYS

As discussed, the type of raceway is often set by local code or by cost factors. On the other hand, the *size* of raceways is governed by NEC requirements. As will be shown, the size of any raceway type is set by the number of conductors, the size of the conductors, and the type of insulation on the conductors. Of course, for cables (either metal-clad or nonmetallic) where the conductors are manufactured as one piece with the outer shield, there is no need to calculate the raceway size. Likewise, the manufacturers of special-purpose wireways specify the number of conductors to be used in the particular wireway. So the calculations for determining raceway size apply only to rigid conduit and EMT.

2-3.1 Calculating raceway size

Figure 2-4 shows the maximum number of conductors permitted by the NEC (for new work) to be used in various trade sizes of conduit tubing. The illustration is simple to use. For example, assume that it is necessary to use three No. 8 AWG conductors in a particular raceway. Find the smallest size of raceway permitted.

Starting with the No. 8 AWG size, move vertically down the row until the first number 3 is found. In this case, the first number 3 is found in the $\frac{3}{4}$-in. size. Thus the smallest conduit or tubing permitted for three No. 8 AWG conductors is $\frac{3}{4}$ in.

Note that four conductors of the same size can be used if the very thin type of THWN nylon insulation is used on the conductors. (In later sections of this chapter we shall discuss the types of insulation in greater detail.)

From a practical standpoint it is often convenient to use raceways that are larger than required. This permits a greater number of conductors, or conductors of a larger size, to be used at a later date (for remodeling or whatever).

Conduit size (inches)	Conductor size																							
	14	12	10	8	6	4	3	2	1	0	2/0	3/0	4/0	250	300	350	400	500	600	700	750	800	900	1000
1/2	4	3	1	1	1	1																		
THWN	8	6	4	2	1	1																		
3/4	6	5	4	3	1	1	1	1	1															
THWN	15	11	7	4	2	1	1	1	1															
1	10	8	7	4	3	1	1	1	1	1	1	1												
THWN	24	18	11	6	4	2	2	1	1	1	1													
1-1/4	18	15	13	7	4	3	3	3	1	1	1	1	1	1	1									
THWN	43	32	20	11	7	4	3	3	2	2	1	1	1	1	1	1								
1-1/2	25	21	17	10	6	5	4	3	3	2	1	1	1	1	1	1	1	1						
THWN	58	43	27	16	9	6	5	4	3	2	2	1	1	1	1	1	1	1						
2	41	34	29	17	10	8	7	6	4	4	3	3	2	1	1	1	1	1	1	1	1	1	1	1
THWN	96	71	45	26	16	9	8	7	5	4	3	3	2	2	1	1	1	1	1	1	1	1	1	1
2-1/2	58	50	41	25	15	12	10	9	7	6	5	4	3	3	1	1	1	1	1	1	1	1	1	1
THWN	137	102	65	37	23	14	12	10	7	6	5	4	3	3	3	2	2	1	1	1	1	1	1	1
3	90	76	64	38	23	18	16	14	10	9	8	7	6	5	4	3	3	3	1	1	1	1	1	1
THWN	—	158	100	58	35	21	18	15	11	9	8	7	6	5	4	3	3	3	2	2	1	1	1	1
3-1/2	121	103	86	52	32	24	21	19	14	12	11	9	8	6	5	5	4	4	3	3	3	2	1	1
THWN	—	134	78	47	29	24	20	15	13	11	9	8	6	5	5	4	4	3	3	3	2	2	1	
4	155	132	110	67	41	31	28	24	18	16	14	12	10	8	7	6	6	5	4	3	3	3	3	3
THWN	—	172	100	61	37	31	26	20	16	14	12	10	8	7	6	6	5	4	3	3	3	3	3	
4-1/2	197	168	140	85	52	40	35	31	23	20	18	15	13	11	9	8	7	6	5	4	4	4	4	3
THWN	—	127	78	48	40	34	25	21	18	15	13	11	9	8	7	6	5	4	4	4	4	3		
5	—	173	105	64	49	44	38	29	25	22	19	16	13	11	10	9	8	6	6	5	5	4	4	
THWN	—	157	96	59	50	42	31	26	22	19	16	13	11	10	9	8	6	6	5	5	4	4		
6	—	152	93	72	63	55	42	37	32	27	23	19	16	15	13	11	9	8	8	7	7	6		
THWN	—	139	85	72	61	45	38	32	27	23	19	16	15	13	11	9	8	8	7	7	6			

FIGURE 2-4   NEC requirements for new work, maximum number of conductors in conduit.

For example, assume that 1-in. conduit is used instead of the $\frac{3}{4}$-in. conduit described in the previous example. With 1-in. conduit, four No. 8 AWG conductors could be used instead of three. This will permit conversion to a four-wire system (for three-phase) from a standard three-wire Edison system. Also, with 1-in. conduit, the conductor size could be increased to No. 6 AWG. However, only three No. 6 AWG conductors are permitted (unless insulation is THWN).

2-4 CONDUCTOR BASICS

Almost any metal will act as a conductor for electrical current. Gold and silver make excellent conductors. However, because of their cost, they are not practical for electrical work. Most wires (and bus bars) used in practical work are made of *copper* or *aluminum*. All metals also have some resistance to the flow of electrical current.

Resistance is the reciprocal of conductance. That is, a metal or material with low resistance has high conductance, and vice versa.

2-4.1 Units of resistance and conductance

The unit of resistance is the *ohm* (after George S. Ohm), whereas the unit of conductance is the *mho* (ohm spelled backward). One ampere of current will flow when there is 1 V across 1 ohm (Ω).

The *resistivity* of a conductor is defined as the resistance (in ohms) of a specific example of a given cross-sectional area and length, and usually at a given temperature. In practical electrical work the resistivity is specified as a *circular mil foot* at 20°C. For example, as shown in Fig. 2-5, the resistivity of copper is 10.4 at 20°C. This means that 1 ft of copper wire 1 mil in diameter will produce 10.4 Ω of resistance at a temperature of 20°C (68°F).

Area in cmils
or
diameter in mils squared

$$R = r\frac{L}{d^2} \quad \text{or} \quad R = r\frac{L}{A}$$

R = Resistance of wire
r = Specific resistance of material
L = Length of wire in feet
d = Diameter in mils
A = Area (in cmils)

Material	Resistivity at 20°C (68 °F) (Ω – cmil / foot)
Silver	9.8
Copper	10.4
Aluminum	17
Tungsten	33
Nickel	50
Iron	60
Manganin	290
Nichrome	660

FIGURE 2-5   Resistance of wire.

2-4.2 Circular mils

The circular mil (or *cmil*) is chosen for electrical work since most conductors or wires are round (and small). A mil is one thousandth of an inch (0.001 in.). A circular mil is a circle 0.001 in. in diameter.

As shown in Fig. 2-5, to find the area of a round conductor in circular mils, simply square the diameter. For example, if the diameter of a wire is 0.050 in., or 50 mils, the area in cmils is 2500 (50 X 50).

Most of the tables and charts found in this book (and those used in the NEC and most electrical equipment manufacturers) use circular mils to describe wire sizes. However, there may be rare cases where you must convert cmils to a square measure, or vice versa. This can be done simply by use of the following equations:

$$\text{cmils} = \text{square mils} \times 1.273$$
$$\text{square mils} = \text{cmils} \times 0.7854$$

For example, assume that a bus bar is $\frac{1}{4}$ in. by 2 in. The square-mil area is found as follows:

$$\frac{1}{4}\text{ in. } = 250 \text{ mils}$$
$$2 \text{ in. } = 2000 \text{ mils}$$
$$250 \times 2000 = 500{,}000 \text{ square mils}$$
$$\text{cmils} = 500{,}000 \times 1.273 = 636{,}500$$

2-4.3 Calculating resistance of conductors

In designing electrical wiring, it is often necessary to find the resistance of conductors or wires. For example, many design problems require that the voltage drop produced by a given length of wiring be known. If the resistance for that length of wiring is known, the voltage drop can be calculated for a given current flow or power consumption.

The resistance of a conductor varies inversely with the cross-sectional area and directly with length. That is, conductor resistance increases as the length increases and decreases as the cross-sectional area increases (the wire size increases). This is shown by the equations of Fig. 2-5.

For example, find the resistance of 1000 ft of aluminum wire with a diameter of 0.050 in, at a temperature of 20°C, using the equations and data of Fig. 2-5. (Note that 0.050 in. is the approximate diameter of No. 16 AWG wire.) D = 0.050 in. or 50 mils; L = 1000 ft; resistivity = 17:

$$\frac{17 \times 1000}{50^2} = \frac{17,000}{2500} = 6.8 \ \Omega$$

2-4.4 Conductor (wire) sizes

The sizes of wires (or conductors) used in practical electrical work conform to the American Wire Gauge (AWG) system. Under the AWG system the smallest wire is No. 40, which has a diameter of 3.145 mils (much too small for electrical wiring but used in electronics). The AWG numbers then decrease (one number at a time) as the wire size increases, up to No. 0000 AWG (also described as 4/0), which is almost $\frac{1}{2}$ in. in diameter (460 mils, to be exact). Wires larger than No. 0000 AWG are described in circular-mil area. To simplify the listing of the larger wire sizes, the wire is described in thousands of circular mils (or MCM). For example, the next size larger than No. 0000 AWG is 250 MCM, which means 250,000 circular mils or 500 mils (exactly $\frac{1}{2}$ in.) in diameter.

The wire-size system is shown in Fig. 2-6, which lists data for copper and aluminum wires in sizes from No. 18 AWG up to MCM 1000 (1 in. in diameter). These are the sizes used in most practical electrical wiring systems. Note that Fig. 2-6 gives the area in cmils and the diameter of each wire in mils. The diameter can be converted to inches by moving the decimal point three places to the left. For example, the diameter of No. 16 AWG is 50.8 mils, which is 0.0508 in. Also note that wire sizes No. 8 through No. 18 AWG are generally solid wires, whereas No. 6 AWG and larger are stranded wires. The diameter column of Fig. 2-6 lists the diameter of *each strand* in stranded wires. This diameter must be squared, and then multiplied by the number of strands, to find the area in cmils. Of course, the area column lists the cmil area for each wire size, so this need not be calculated. The diameter and number of wire strands are given only for reference.

Figure 2-6 lists the resistance of both copper and aluminum conductors for each wire size. A glance at the resistance columns brings up several points to be remembered in practical electrical wiring. Obviously, the larger the wire, the less resistance—all other factors (length, temperature, material) being equal. Next, aluminum has

Size (AWG or MCM)	Area	Number of wires	Diameter each wire	dc resistance at 25 °C (77 °F) (Ω / 1000 feet)	
				Copper	Aluminum
AWG					
18	1,620	Solid	0.0403	6.51	10.7
16	2,580	Solid	0.0508	4.10	6.72
14	4,110	Solid	0.0641	2.57	4.22
12	6,530	Solid	0.0808	1.62	2.66
10	10,380	Solid	0.1019	1.018	1.67
8	16,510	Solid	0.1285	0.6404	1.05
6	26,240	7	0.0612	0.410	0.674
4	41,740	7	0.0772	0.259	0.424
3	52,620	7	0.0867	0.205	0.336
2	66,360	7	0.0974	0.162	0.266
1	83,690	19	0.0664	0.129	0.211
0	105,600	19	0.0745	0.102	0.168
00	133,100	19	0.0837	0.0811	0.133
000	167,800	19	0.0940	0.0642	0.105
0000	211,600	19	0.1055	0.0509	0.0836
MCM					
250	250,000	37	0.0822	0.0431	0.0708
300	300,000	37	0.0900	0.0360	0.0590
350	350,000	37	0.0973	0.0308	0.0505
400	400,000	37	0.1040	0.0270	0.0442
500	500,000	37	0.1162	0.0216	0.0354
600	600,000	61	0.0992	0.0180	0.0295
700	700,000	61	0.1071	0.0154	0.0253
750	750,000	61	0.1109	0.0144	0.0236
800	800,000	61	0.1145	0.0135	0.0221
900	900,000	61	0.1215	0.0120	0.0197
1000	1,000,000	61	0.1280	0.0108	0.0177

FIGURE 2-6   Conductor resistance in ohms per 1000 feet.

more resistance than copper, all other factors being equal. *As a guideline or rule of thumb, it is necessary to use the next larger wire size when aluminum is used instead of copper.* For example, the resistance for 1000 ft of aluminum No. 16 AWG is approximately the same as for 1000 ft of copper No. 18 AWG. This is a guideline only, however, and applies primarily to the smaller wire sizes found in house and small-building wiring.

The resistance columns give the resistance for 1000 ft of conductor at a temperature of 25°C (77°F). This is different from the 20°C temperature of Fig. 2-5. If you bother to calculate the resistance using both Figs. 2-5 and 2-6, you will note a slight difference in resistance (resulting from the small difference in the two temperatures). However, if there is a large difference in temperature, there will be a corresponding difference in resistance (all other factors remaining the same). This is discussed next.

2-4.5 Conductor resistance as affected by temperature

The resistance of any conductor will vary with temperature. For conductors used in practical electrical wiring, the resistance increases when temperature increases. Typically, the resistance of conductors is listed at temperatures in the range of 20 to $25°C$ (68 to $77°F$). This range is considered the normal or average operating temperature of electrical wiring. The actual operating temperature of wiring is controlled by two major factors. First, there is the ambient temperature, or temperature that surrounds the wiring. Second, there is the temperature that results from heat caused by current flowing in the wires. As current increases, the generated heat increases, as does the resistance. (If the opposite were true, the wiring would probably burn out.)

In theory, if you lower the temperature of any conductor far enough, the resistance will be zero. In practical work, this is impossible. For one reason, if there is any current flow, there will be some heat and an increase in temperature (thus producing an increase in resistance). However, there is an *inferred zero resistance temperature* for conductors, shown in Fig. 2-7 for several common conductors together with equations for resistance at different tem-

$$R_1 = R_2 \frac{IZR + T_1}{IZR + T_2} \qquad R_2 = R_1 \frac{IZR + T_2}{IZR + T_1}$$

R_1 = resistance at low temperature

R_2 = resistance at high temperature

T_1 = low temperature

T_2 = high temperature

IZR = inferred zero resistance temperature

Inferred temperature for zero resistance ($C°$)	Material
-243	Silver
-234.5	Copper
-236	Aluminum
-202	Tungsten
-147	Nickel
-180	Iron
-6250	Nichrome

FIGURE 2-7   Determining conductor resistance at different temperatures.

peratures. To use the data in Fig. 2-7 it is necessary to know the resistance of the conductor at a particular temperature. Armed with this information you can find the resistance of the conductor at any other temperature.

For example, Fig. 2-6 shows a resistance of 4.1 Ω for 1000 ft of No. 16 AWG copper wire at 25°C (77°F). Assume that the temperature increases to 50°C (122°F). Find the resistance for the same wire using the equations of Fig. 2-7.

$$\text{resistance at } 50°C = 4.1 \times \frac{234.5 + 50}{234.5 + 25} = 4.5 \text{ } \Omega \text{ (approximately)}$$

Now assume that the temperature drops to 0°C (32°F). Find the resistance for the same wire.

$$\text{resistance at } 0°C = 4.1 \times \frac{234.5 + 0}{234.5 + 25} = 3.8 \text{ } \Omega \text{ (approximately)}$$

2-4.6 Skin effect

Thus far we have discussed the resistance of conductors in reference to direct current (dc). In practical electrical wiring problems, we must deal with alternating current (ac). This brings up the problem of *skin effect*. When alternating current flows in a conductor, the current tends to flow on the outside, or "skin," of the conductor. This decreases the conductor area and increases resistance. That is, the ac resistance of a conductor is higher than the dc resistance, all other factors being equal. As the frequency of the alternating current increases, the skin effect becomes more pronounced and the resistance increases.

From a practical standpoint, skin effect is a problem only on larger conductors (typically on conductors of No. 1 AWG or larger). Also, the skin effect is more pronounced when conductors are enclosed in metal (such as in metal-clad cable, rigid conduit, etc.).

Figure 2-8 shows multiplying factors for converting dc resistance to ac resistance (at the commonly used frequency of 60 Hz) for both copper and aluminum. Note that skin effect is more pronounced for copper than for aluminum.

To use Fig. 2-8 multiply the dc resistance by the multiplying factor. For example, Fig. 2-6 shows that the resistance for 1000 ft of MCM 800 copper conductor is 0.0135 Ω. Assume that this same

Size AWG or MCM	Conductors in metal cables or raceways		Conductors in air or non–metal cables and raceways	
	Copper	Aluminum	Copper	Aluminum
AWG				
3	1	1	1	1
2	1.01	1	1	1
1	1.01	1	1	1
0	1.02	1	1.001	1
00	1.03	1	1.001	1.001
000	1.04	1.01	1.002	1.001
0000	1.05	1.01	1.004	1.002
MCM				
250	1.06	1.02	1.005	1.002
300	1.07	1.02	1.006	1.003
350	1.08	1.03	1.009	1.004
400	1.10	1.04	1.011	1.005
500	1.13	1.06	1.018	1.007
600	1.16	1.08	1.025	1.010
700	1.19	1.11	1.034	1.013
750	1.21	1.12	1.039	1.017
800	1.22	1.14	1.044	1.017
900	1.27	1.17	1.056	1.022
1000	1.30	1.19	1.067	1.026

DC Resistance • Multiplier = AC Resistance

FIGURE 2-8   Skin effect multiplying factors.

1000 ft of copper conductor is used with 60-Hz alternating current, enclosed in a metal raceway. Find the ac resistance.

$$\text{dc resistance} \times \text{multiplier} = \text{ac resistance}$$
$$0.0135 \times 1.22 = 0.01647 \ \Omega$$

2-5 CONDUCTOR MATERIALS AND INSULATION

Copper or aluminum are used as conductors for almost all modern electrical wiring. Of the two, copper is most frequently used for all sizes of wire and is used almost exclusively for the small wire sizes. Each conductor material has its advantages.

Copper has less resistance (10.4 Ω/cmil ft, compared to 17 Ω/cmil ft for aluminum) and is stronger. Copper will thus carry more current for a given wire size.

Aluminum is less expensive and is lighter (about one-third lighter than copper). Thus a larger aluminum conductor could weigh less (and cost less) than a smaller copper conductor of the same current capacity.

No matter what material is used, all conductors must have some form of insulation. The type of insulating material is determined by the conditions or environment in which the conductor is to be used. Heat and moisture are the main problems. Excessive heat (caused either by external conditions or high currents, or both) can melt insulation; extreme heat can burn insulation. Excessive moisture can penetrate some insulation and result in a short circuit. In addition, age can cause insulation to deteriorate. For example, age and heat will cause natural rubber to crack. The amount of insulation is determined by the voltage between conductors. A higher voltage requires more insulation.

The NEC has classifications for types of insulating materials and for voltages. There are six voltage classifications for insulation: 600, 1000, 2000, 3000, 4000, and 5000 V. The NEC requires that conductors be identified by the first two numbers stamped or printed on the insulation. For example, the numbers 10 indicate that the insulation is suitable for voltages up to 1000 V *maximum.* When there is no identifying number, the conductor insulation is for use with 600 V *maximum.*

There are several classifications of insulation types in general use. Each type is assigned an identifying letter. The letters indicate insulation material or application or both. There are five letters for the type classifications: R for rubber, T for thermoplastic, N for nylon, H for heat resistant, and W for moisture resistant. The basic NEC classifications are as follows:

Type	Material and Characteristics	Application	Maximum Operating Temp. (°F)
R	Rubber	Dry locations	140
RH	Heat-resistant rubber	Dry locations	167
RHH	Higher-temperature heat-resistant rubber	Dry locations	194
RHW	Heat- and moisture-resistant rubber	Dry or wet locations	167
T	Thermoplastic	Dry locations	140
TH	Heat-resistant thermoplastic	Dry locations	167
THW	Heat- and moisture-resistant thermoplastic	Dry or wet locations	167

Type	Material and Characteristics	Application	Maximum Operating Temp. (°F)
THWN*	Heat- and moisture-resistant thermoplastic with nylon covering	Dry and wet locations	167

*Note that THWN is thinner than other insulations in general use. Thus the overall size of a THWN conductor (including the insulation) is smaller than conductors with other insulating materials (of the same size number).

2-6 CONDUCTOR CURRENT CAPACITY (AMPACITY)

The word *ampacity* means capacity in amperes. When applied to conductors, the term "ampacity" means the *maximum* current, in amperes, that can be carried in a conductor, *with safety*. The NEC provides tables that specify the current-carrying capacity, or ampacity, of commonly used conductor sizes, under a given set of conditions. Such tables are reproduced in Figs. 2-9 and 2-10.

	Size AWG or MCM	Copper R,T,TW, 60° C (140° F)	RH,RHW,TH THWN,THW 75°C (167° F)	Aluminum R,T,TW 60° C (140° F)	RHW,TH THW,THWN 75°C (167°F)
AWG	14	20	20	—	—
	12	25	25	20	20
	10	40	40	30	30
	8	55	65	45	55
	6	80	95	60	75
	4	105	125	80	100
	3	120	145	95	115
	2	140	170	110	135
	1	165	195	130	155
	0	195	230	150	180
	00	225	265	175	210
	000	260	310	200	240
	0000	300	360	230	280
MCM	250	340	405	265	315
	300	375	445	290	350
	350	420	505	330	395
	400	455	545	355	425
	500	515	620	405	485
	600	575	690	455	545
	700	630	755	500	595
	750	655	785	515	620
	800	680	815	535	645
	900	730	870	580	700
	1000	780	935	625	750

FIGURE 2-9   Ampacity of single, insulated conductors in free air.

Size AWG or MCM	Copper R,T,TW 60° C(140°F)	RH,RHW,TH THW,THWN 75°C(167°F)	Aluminum R,T,TW 60° C(140° F)	RH,RHW,TH, THW,THWN 75° C(167°F)
AWG				
14	15	15	—	—
12	20	20	15	15
10	30	30	25	25
8	40	45	30	40
6	55	65	40	50
4	70	85	55	65
3	80	100	65	75
2	95	115	75	90
1	110	130	85	100
0	125	150	100	120
00	145	175	115	135
000	165	200	130	155
0000	195	230	155	180
MCM				
250	215	255	170	205
300	240	285	190	230
350	260	310	210	250
400	280	335	225	270
500	320	380	260	310
600	355	420	285	340
700	385	460	310	375
750	400	475	320	385
800	410	490	330	395
900	435	520	355	425
1000	455	545	375	445

FIGURE 2-10   Ampacity of three (or less) conductors in a cable or raceway.

Figure 2-9 shows the ampacity of single insulated conductors in free air. Note that the ampacity is set by the type of insulation and the conductor material, as well as by the conductor size. For example, for No. 8 AWG copper conductors, the ampacity is 55 A, when R, T, or TW insulations are used, but is increased to 65 A for RH, RHW, TH, and THW insulations. Using the same No. 8 AWG size and an aluminum conductor, the ampacity ratings are reduced to 45 and 55 A.

The information of Fig. 2-9 is not too realistic in that most electrical wiring does not involve a single conductor in free air. Typically, three (or at least two) conductors are used in cables or raceways.

Figure 2-10 shows the ampacity of three (or less) conductors in a cable or raceway. Again, the ampacity is set by insulation, conductor material, and size. However, the ampacity of conductors in cables or raceways is always less than for a single conductor in free air. For example, using the same No. 8 AWG size, the ampacity for copper conductors is reduced to 40 and 45 A.

2-6.1 Temperature derating factors

Although the information in Fig. 2-10 can be used for practical work, it has certain limitations. The data in Fig. 2-10 are based on operating the electrical wiring in an ambient temperature not over 30°C (86°F). If the normal ambient temperature is higher than 30°C, the conductor ampacity must be *derated*. That is, a lower current must be used for reasons of safety.

Conductor derating factor		Elevated ambient temperature
R, T, TW	RH, RHW, TH THW, THWN	
0.82 (1.22)	0.88 (1.14)	40°C (104°F)
0.71 (1.41)	0.82 (1.22)	45°C (113°F)
0.58 (1.73)	0.75 (1.33)	50°C (122°F)
0.41 (2.44)	0.67 (1.5)	55°C (131°F)
—	0.58 (1.73)	60°C (140°F)
—	0.35 (2.86)	70°C (158°F)
—	0.20 (5.0)	75°C (167°F)

FIGURE 2-11   Conductor derating factors for elevated temperatures.

Figure 2-11 shows derating factors for ampacity of conductors used in ambient temperature over 30°C (86°F). Note that the derating factor for conductors with insulations normally used in lower temperatures is different than that for high-temperature insulations. For example, assuming an ambient temperature of 40°C (104°F), the derating factor is 0.82 for R, T, and TW insulations, and 0.88 for RH, RHW, TH, THW, and THWN insulations.

To find the ampacity of a conductor operating at an elevated temperature (above 30°C), *multiply* the 30°C ampacity (Fig. 2-10) by the derating factor. For example, as shown in Fig. 2-10, the ampacity of No. 10 AWG copper conductor with R, T, or TW insulation is 30 A. If this same conductor is operated at 55°C (131°F), the ampacity is reduced to 12.3 A (30 × 0.41).

Keep in mind that in practical electrical work the problem of ampacity derating factors will generally be stated in reverse. That is, a given load will require a given current. If the wiring is normally operated at temperatures above 30°C, a large wire size or a different insulation will be required. This means that you must *divide* the required current by the derating factor to find a "derated current," and then use this higher derated current to find the correct conductor size and insulation type.

For example, assume that the load requires 100 A and that three TW copper conductors are to be operated in a cable at ambient temperatures of 55°C (131°F).

The first step is to divide the 100 A by the TW derating factor for 55°C (131°F), which is 0.41. As an alternative multiply by the reciprocal of the derating factor (the number in parentheses). Since it is easier to multiply than divide, use the reciprocal (of 2.44 in this case) to find a derated current of 244 A (100 A × 2.44).

Next, check the TW column for copper conductors in Fig. 2-10 to find the conductor size that will accommodate 244 A. Always use the *next larger size*, unless an exact size can be found. In this case, 350 MCM will accommodate 260 A (for copper with TW insulation). A size of 300 MCM is slightly low (for 240 A), so the 350 MCM should be used.

As an alternative, practical solution, if the insulation can be changed to THW, then the size can be reduced to No. 0 AWG. This drastic reduction in size is possible because of the different derating factor of THW. As shown in Fig. 2-11, the derating factor for THW insulation at 55°C (131°F) is 0.67, and the reciprocal is 1.5. Using the reciprocal, multiply the required current of 100 A by 1.5 to find a derated current of 150 A. Then check the THW column for copper conductors in Fig. 2-10 to find the conductor size that will accommodate 150 A. In this case, No. 0 AWG will do the job.

2-6.2 Multiple conductor derating factors

The information in Fig. 2-10 is based on using a maximum of three conductors in a raceway or cable. This applies to the great majority of practical electrical work (most houses, small buildings, etc.). However, in larger electrical systems, it is often convenient (and economical) to have many conductors in a raceway. When more than three conductors are used in the same raceway or cable, the conductor ampacity must be derated, so that a large conductor is required for safety.

Figure 2-12 shows derating factors for ampacity of multiple conductors (more than three) used in the same raceway or cable. To find the ampacity of a *single conductor* operating in a raceway or cable with more than three conductors, *multiply* the single conductor ampacity (Fig. 2-9) by the derating factor. For example, as shown in Fig. 2-9, the ampacity of No. 10 AWG copper conductor with R, T,

Conductor derating factor	Number of conductors in raceway
0.8 (1.25)	4 to 6
0.7 (1.43)	7 to 24
0.6 (1.67)	25 to 42
0.5 (2.0)	43 or more

FIGURE 2-12   Conductor derating factors for more than three conductors in a raceway.

or TW insulation is 40 A. If this same conductor is operated in a raceway with four to six conductors, the ampacity is reduced to 32 A (40 × 0.8).

To find the ampacity of a three-wire system operating in a raceway or cable with more than three conductors, *multiply* the three-conductor ampacity (Fig. 2-10) by the derating factor. For example, as shown in Fig. 2-10, the ampacity of No. 10 AWG copper conductor with R, T, or TW insulation in a three-conductor system is 30 A. If these same conductors are operated in a raceway with four to six conductors, the ampacity is reduced to 24 A (30 × 0.8).

For practical work the problem of multiple conductor derating will be stated in reverse. A given load requires a given current, and a larger wire size is required when multiple conductors are used.

For example, assume that the load requires 100 A and that three TW copper conductors are to be operated in a cable with a total four to six conductors.

The first step is to divide the 100 A by the four to six conductor derating factor (0.8), or multiply by the reciprocal (1.25). Using the reciprocal, multiply the required current of 100 A by 1.25 to find a derated current of 125 A. Then check the TW column for copper conductors in Fig. 2-10 to find the conductor size that will accommodate 125 A. In this case No. 0 AWG will do the job.

2-6.3 Three-phase derating factors

When three-phase wiring is used, several special derating factors must be considered.

In three-phase four-wire systems that are balanced, the fourth (or neutral) wire carries no current and produces no heat. Thus balanced three-phase four-wire systems can be rated the same as three-wire system as far as the ampacity of conductors is concerned. Even if the three-phase four-wire systems are unbalanced, the line currents decrease as neutral currents increase; thus, the net heating effect is the same, and the information of Fig. 2-10 applies.

If a three-phase four-wire system is used with fluorescent (discharge) lighting, the neutral conductor current is not zero. This applies even if the three-phase four-wire system is balanced. Thus in any three-phase four-wire system involving fluorescent lights, the wire sizes must be derated in ampacity using the factors of Fig. 2-12.

2-6.4 Neutral wire derating factors

If the current of the neutral wire in a three-phase system (or a single-phase system) is over 200 A, *the NEC allows the size of the neutral wire to be reduced.*

The size of the neutral wire is based on a current equal to 200 A, plus 70 per cent of the current over 200 A. For example, if the maximum current in the neutral wire is 400 A, the current used to find the correct wire size is 340 A (400 A − 200 A = 200 A; 200 A × 70 per cent = 140 A; 200 A + 140 A = 340 A). However, *the NEC does not permit this reduction if any of the load is comprised of fluorescent lighting.*

2-7 CONDUCTOR VOLTAGE DROP

If conductors had no resistance, there would be no voltage difference (or drop) between the source and load, nor would there be power consumption in the conductors. In practical work, since all conductors have some resistance, there is always some voltage drop. Also, since there is some resistance and voltage drop, there is some power consumption in any conductor. Thus all conductors generate some heat.

From a practical wiring standpoint, the main concern is that the voltage drop does not exceed a certain percentage of the source voltage. For example, *the NEC permits a maximum of 3 per cent voltage drop to the farthest outlet in a branch circuit and a 5 per cent voltage drop if feeders are included.* (Feeders and branch circuits are discussed in Chapter 3.)

In a typical two-wire 120-V system, the maximum voltage drop that could be tolerated from the service entrance to the farthest outlet is 6 V (120 × 5 per cent). Keep in mind that this represents a

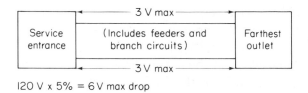

120 V x 5% = 6 V max drop

FIGURE 2-13 Maximum conductor voltage drop in 120V, 2-wire system.

voltage drop of 3 V (or one-half of 6 V) for each of the two conductors, as shown in Fig. 2-13. If each conductor is permitted a 6-V drop, the total drop to the farthest outlet would be 12 V.

For a given wiring design problem, a larger wire size will produce a larger cmil area and a lower resistance. Thus a larger wire size produces a lower voltage drop. A simple solution to any voltage drop problem is to use the largest possible wire size, but this is neither practical nor economical. As any contractor knows, an increase in one or two wire sizes can sometimes raise the overall cost of a job to a point where it is no longer competitive.

A more practical solution is to use a wire size that produces a voltage drop *just below* the maximum permitted. Of course, the wire size selected for voltage drop must also be capable of carrying the maximum rated current.

2-7.1 Finding correct wire size to provide a given voltage drop

Although there are several methods for finding a wire size that will produce a given voltage drop, the following is the most direct. This method involves finding the area *in cmils* that will produce the voltage drop, and choosing the *next largest standard wire size*. Then the selected wire size is checked for ampacity, and derated (if necessary) for heat, number of conductors, skin effect, and so on, as described in Sections 2-4 through 2-6.

The basic equation for finding the cmil area for a given voltage drop is

$$\text{cmil} = \frac{\substack{\text{resistivity} \\ \text{of material}} \times \text{current} \times \text{distance}}{\text{source voltage} \times \text{percentage of drop}}$$

For single phase, the distance is two times the distance from source to load. This applies to both two- and three-wire single-phase systems, since the neutral wire carries no current for balanced loads. If the load is unbalanced the neutral wire will always carry less current than either of the other wires. Thus the wire size selected for the other wires will produce a lower voltage drop in the neutral wire.

For three phase, the distance is 1.732 times the distance from source to load.

In practical problems the load may be expressed in terms of power (watts or kilowatts) rather than in current (amperes). For single phase, convert power to current by the formula $I = P/E$, using the *maximum load* in watts for P and the source voltage for E. For three phase, the conversion is more complex: $I = P/E \times 1.732 \times$ power factor.

Example of single-phase voltage drop. Assuming a resistivity of 17 for aluminum, what size aluminum conductors are required to supply a 100-A load 300 ft from a 120-V source with a voltage drop not to exceed 3 per cent?

$$\text{cmil} = \frac{17 \times 100 \times (300 \times 2)}{120 \times 0.03} = 283{,}333 = \text{approximately 284 MCM}$$

The next largest standard size is 300 MCM (Fig. 2-6).

Always check the selected standard size for correct ampacity. As shown in Fig. 2-10, 300-MCM aluminum conductors with any type of insulation will carry more than 100 A. Of course, if the conductors are to be operated at high temperatures, or with many conductors in a raceway, the derating factors must be applied. Since conductors of larger size are involved, it may be well to check the skin effect (Section 2-4.6).

The maximum voltage drop is 3.6 V with 100 A of current. This results in a maximum dc resistance of 0.036 Ω ($R = E/I$; 3.6/100 = 0.036). The total length of the conductor is 600 ft (300 ft from source to load and 300 ft from load back to source). Thus the maximum allowable resistance for 600 ft of conductor is 0.036 Ω. Using the information of Section 2-4.3, the direct current resistance for 600 ft of 300-MCM conductor with an assumed resistivity of 17 is

$$\frac{17 \times 600}{300{,}000} = \frac{10{,}200}{300{,}000} = 0.034$$

Since the actual dc resistance of 0.034 is less than the desired maximum of 0.036, 300 MCM will be satisfactory. However, as shown in Fig. 2-8, a multiplying factor of 1.02 must be applied to aluminum 300-MCM conductors in a metal raceway, for skin effect. This makes the ac resistance 0.03468 (0.034 × 1.02). The ac resistance is lower than the 0.036 maximum (and will produce a voltage drop of 3.468 V).

Of course, this example is purposely close. In practical work, the 300-MCM conductors can probably be used. If not, the next larger standard size of 350 MCM provides more than enough tolerance in resistance.

Example of three-phase voltage drop. The power for the single-phase example is approximately 11.5 kW. Now assume that a 12-kW load, the same 300 ft from the source, is to be supplied with a three-phase 240-V source at a power factor of 0.8. The voltage drop is still not to exceed 3.6 V (which is 1.5 per cent of 240 V), and aluminum conductors must be used.

First convert the 12-kW load to a current:

$$I = \frac{12{,}000}{240 \times 1.732 \times 0.8} = \text{approximately 36 A}$$

With a current of 36 A, find the cmil area:

$$\text{cmil} = \frac{17 \times 36 \times (300 \times 1.732)}{240 \times 0.015} = \text{approximately 88{,}332}$$

The next larger standard size is No. 0 AWG (Fig. 2-6).

Check this standard size for correct ampacity. As shown in Fig. 2-10, No. 0 AWG aluminum conductors with any type of insulation will carry more than 36 A. Check the derating factors as described in Sections 2-4 through 2-6, paying particular attention to Section 2-6.3. Skin effect can be ignored, because the skin effect multiplying factor (Fig. 2-8) for No. 0 AWG aluminum conductor is 1.

Chapter 3

Feeders, Branch Circuits, and Overcurrent Protection

In this chapter we shall discuss distribution of electrical power from the service entrance to the electrical outlets. In all electrical systems, this distribution must include overcurrent protection. In all but the simplest systems, the distribution must also include both feeders and branch circuits.

The need for overcurrent protection in all electrical circuits is obvious. Without a fuse or circuit breaker at some point in the circuit, a short in the wiring or in an appliance will cause the conductors to overheat. This will destroy the conductors and cause a fire. The need for feeders and branch circuits will become obvious when you study the design alternatives.

One design alternative is to run a common two-wire system throughout a building and tap off as many outlets and permanent fixtures as needed. The common system can then be provided with overcurrent protection at the service entrance. This overcurrent protection must have an ampacity equal to the total of all the loads. With such a design, individual appliances could be destroyed by too much current, without actuating the overcurrent device. On the other hand, if the overcurrent ampacity is lowered to protect the individual appliances, the overcurrent device will be actuated by normal loads when all the loads are used simultaneously.

Another design alternative is to use a separate two-wire system for each outlet and fixture and to provide a separate overcurrent device for each two-wire system. Obviously this is not economical, nor is it practical.

3-1 BASIC FEEDER AND BRANCH CIRCUIT DESIGN

The problems created by either of these extreme design alternatives can be solved with feeder and branch circuits, which are approved, and required, by NEC. The feeder and branch circuit relationship can best be understood by reference to Figs. 3-1 and 3-2, which show the elementary wiring design for a seven-unit apartment building.

As shown, the service entrance is supplied power through a typical three-wire system (Chapter 1) from the utility company. One

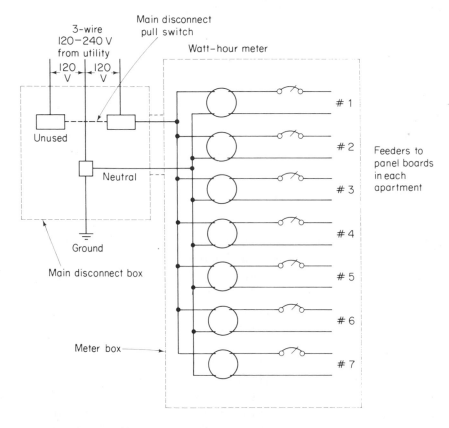

FIGURE 3-1   Service entrance of elementary wiring design for 7-unit apartment building.

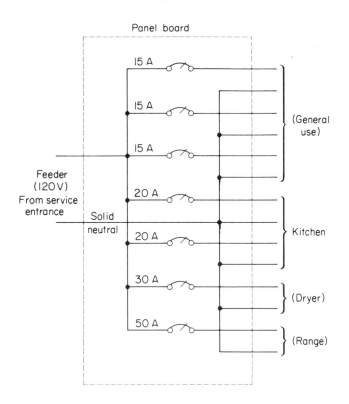

FIGURE 3-2   Elementary wiring design for panel board of individual apartment.

of the three wires is unused. One of the two active wires is grounded at the service entrance main disconnect (or main switch) box. (This is the *neutral* wire, as discussed in Chapter 4.)

From this point, seven feeders supply power to the corresponding seven apartments. Each feeder has a separate meter and overcurrent device (a circuit breaker in this case). The meter and circuit breaker are located in an enclosure at the service entrance, next to the main disconnect box. Each feeder terminates at a panelboard located within the corresponding apartment.

The panelboards are provided with seven circuit breakers, one for each of seven branch circuits to outlets, appliances, and so on. Note that the ampacity of all circuit breakers is not the same. Instead, there are three 15-A, two 20-A, one 30-A, and one 50-A branch circuits, with corresponding circuit breakers. As discussed later in this chapter, *the NEC allows branch circuits of five different ampacities (15, 20, 30, 40, and 50 A).*

3-1.1 Feeder-branch circuit design problems

The main steps in the design of feeder and branch circuits are as follows:

1. The selection of conductors of the correct size to provide the necessary ampacity.
2. The selection of conductor size to produce a minimum voltage drop. (Although the NEC permits a maximum voltage drop of 5 per cent from service entrance to final outlet, this percentage is not necessarily efficient.)
3. The selection of overcurrent devices to match the feeder and conductor ampacities.
4. Finally, and probably most important, the division or arrangement of the loads among the feeders and branch circuits so that there is no overload on any one circuit but each circuit is used to its full capacity. That is, the circuits should be designed so that there is no unnecessary expense caused by an ampacity far beyond that needed for any given circuit.

Each of these problems is discussed in the following sections.

3-2 CALCULATING FEEDER LOADS

The first step in solving any wiring design problem is to calculate the load or loads. With the load established, and a given source voltage, the required ampacity can be calculated. Then the wire size can be matched to provide the necessary ampacity (with minimum voltage drop), all within the framework of NEC requirements.

A simple way to find the ampacity for a feeder would be to add up the maximum ampacity for all branch circuits supplied by the feeder. For example, the ampacity for each feeder in Fig. 3-1 is 165 A (3 × 15 A = 45 A; 2 × 20 A = 40 A; 45 + 40 + 30 + 50 = 165). But whereas this method of calculation could apply to the feeders in Fig. 3-1, it would not be adequate in all situations.

The NEC permits the feeder load to be calculated on the basis of the actual total branch circuit load, in amperes. However, the feeder must have an ampacity to provide for the *maximum anticipated load.*

3-2.1 Demand factor ratings

The maximum anticipated load provision by the NEC is based on the fact that all appliances are not used simultaneously nor at their full rating. For example, a typical household electric range requires a maximum of about 12 kW. However, this maximum power is consumed only when all burners, the broiler, and the oven are operated simultaneously. Since household ranges are rarely used this way, the NEC permits the power consumption for a single 12-kW range to be calculated on the basis of 8 kW. The difference between the two power ratings is called the *demand factor* or *demand load.*

The demand factor can be applied to portions of the load for such typical residential equipment as lighting (or illumination), small appliances, clothes dryers, and electric ranges. In general, the demand factor cannot be applied to such factors as heating or air conditioning.

Figures 3-3, 3-4, and 3-5 show the NEC demand factors for clothes dryers, household ranges, and lighting and small appliance loads, respectively. The demand factor is expressed differently in each of these figures. In Fig. 3-3 the demand factor for household electric clothes dryers is expressed as a per cent of total power rating for all the dryers. For example, if one feeder must supply seven dryers, and each dryer requires 1000 watts (1 kW), the total power

Number of dryers in system	Demand factor (percentage)
1 to 4	100
5	80
6	70
7	65
8	60
9	55
10	50
11 to 13	45
14 to 19	40
20 to 24	35
25 to 29	33
30 to 34	30
35 to 39	28
40 or more	25

FIGURE 3-3   NEC demand factors for electric clothes dryers.

Number of ranges in system	Demand factor (kilowatts)
1	8
2	11
3	14
4	17
5	20
6	21
7	22
8	23
9	24
10	25
11	26
12	27
13	28
14	29
15	30
16	31
17	32
18	33
19	34
20	35
21	36
22	37
23	38
24	39
25	40
26 to 40	15 (plus 1 kW for each range)
41 or more	25 (plus $\frac{3}{4}$ kW for each range)

FIGURE 3-4   NEC demand factors for electric ranges (for ranges of 12 kW or less).

rating is 7 kW. As shown in Fig. 3-3, a 65 per cent demand factor is applied, resulting in a reduced power demand of 4.55 kW. This figure of 4.55 kW can be used to calculate the feeder ampacity. Of course, the 4.55-kW figure must be added to any other load supplied by the feeder.

In Fig. 3-4 the demand factor for household electric ranges is expressed as a power (in kW) for the number of ranges involved. For example, if one feeder must supply seven ranges, and each range requires 12 kW (a typical power rating), the total power rating is 22 kW.

In Fig. 3-5 the demand factor for combined lighting and small appliance loads is expressed as a per cent of the combined load, related to the type of building. For example, if the combined load is 18 kW in an apartment or single residence, the demand factor is 100 per cent of the first 3 kW, and 35 per cent for the remaining 15 kW. This results in a reduced power demand of 8.25 kW (15 kW × 35 = 5.25 kW; 5.25 kW + 3 kW = 8.25 kW).

Occupancy	Portion of lighting load to which demand factor applies (kilowatts)	Demand factor (percentage)
Hotels, motels and apartments (without cooking facilities for tenants)	Up to 20 kW 20 to 100 kW Over 100 kW	50 40 30
All other dwellings	Up to 33 kW 2 to 120 kW Over 120 kW	100 35 25
Hospitals	Up to 50 kW Over 50 kW	40 20
Warehouses (storage)	Up to 12.5 kW Over 12.5 kW	100 50
All others	Total	100

FIGURE 3-5   NEC demand factors for lighting and small appliance loads.

If the same 18-kW load is used in a hotel, in a motel, or in any apartment where cooking facilities are not supplied to the tenants, the reduced power demand is 9 kW. As shown in Fig. 3-5, a 50 per cent demand factor is applied to loads below 20 kW in hotels, motels, and so on (dwellings without individual cooking facilities), and 50 per cent of 18 kW is 9 kW.

3-2.2 Demand factor for residential lighting, small appliances, and laundry facilities

The NEC has special demand factors for residences—either apartments or single-family dwellings. The residential lighting load is considered to be 3 watts (W) per square foot. Thus a 1000-ft^2 apartment is considered to require 3 kW for lighting, whereas a 1500-ft^2 single-family residence requires 4.5 kW.

The small appliance load is considered to be 3 kW, without regard to the size of the residence. Thus both the 1000-ft^2 apartment and the 1500-ft^2 home require 3 kW for small appliances.

The laundry facility load is considered to be 1.5 kW, without regard to the size of the residence. However, in an apartment building where there is a common laundry room and each apartment is not provided with individual laundry facilities, the 1.5-kW load can be eliminated when calculating feeder loads for individual apartments.

Note that the laundry facility load is for the washing machine and not for an electric clothes dryer.

3-2.3 Demand factor for nonresidential lighting

The NEC has demand factors for nonresidential lighting loads. These factors are shown in Fig. 3-6 and apply to all types of buildings, including hotels, motels, and apartments without cooking facilities. As shown in Fig. 3-6, a 1000-ft^2 motel room requires only 2 kW for lighting, compared to 3 kW for a regular apartment with cooking facilities.

3-2.4 Calculating feeder ampacity for apartments

The following is an example of how the NEC demand factors can be applied to find the correct ampacity of feeders used in apartment

Occupancy	Load in watts per square foot
Hotels, motels and apartments (without cooking facilities for tenants)	2
Armories and auditoriums	1
Banks	2
Barber shops and beauty parlors	3
Churches	1
Clubs	2
Court rooms	2
Garages (commercial storage)	0.5
Hospitals	2
Industrial commercial buildings	2
Lodge rooms	1.5
Office buildings	5
Restaurants	2
Schools	3
Stores	3
Warehouses (storage)	0.25

Other dwellings (single—family and apartments)
3 kW per square foot

FIGURE 3-6   NEC lighting load factors.

buildings. Assume that the apartment building consists of 10 identical apartments. Each apartment has 1000 ft^2 and is provided with a 2-kW dishwasher, a 10-kW space heater, and a 12-kW electric range. There is a common laundry room with a 20-kW water heater, a 3-kW washing machine, and a 5-kW dryer. In addition, there are exterior house lights requiring 2 kW. The supply is 120/240 V (generally required when electric ranges are used).

The elementary wiring design is shown in Fig. 3-7. The service entrance equipment consists of the main disconnect box, 11 meters, and 11 circuit breakers. There is a panelboard in each apartment, supplied by a separate feeder. There is a separate panelboard for house power located near, but not at, the service entrance. By adding up the power requirements for each panelboard, we can see that the house feeder requires 30 kW, as do each of the apartment feeders. However, because of the different demand factors, the apartment feeders are not necessarily the same size as the house feeder.

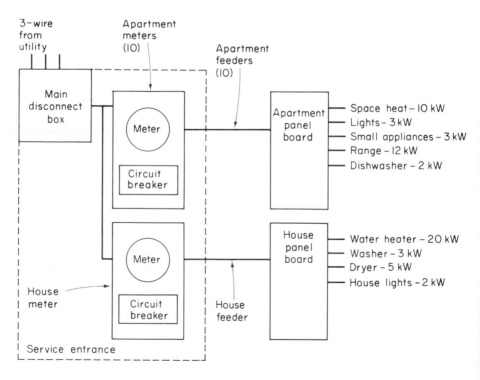

FIGURE 3-7 Elementary wiring diagram showing loads for individual apartments and house loads.

Calculating the house feeder. The demand factor for a water heater, washing machine, house lights, and one dryer is 100 per cent. Thus the total house power is 30 kW. With a 240-V source, the house feeder must be capable of supplying 125 A (30 kW/240 = 125 A). The correct wire size for a three-wire system is given in Fig. 2-10.

Calculating the apartment feeder. The demand factor for a space heater and dishwasher is 100 per cent. Thus the fixed load is

space heater	10 kW
dishwasher	2 kW
	12 kW

The demand factor for a single electric range is considered as 8 kW, even though the range is rated at 12 kW. Thus the load for each apartment, excluding the lighting and small appliances, is

space heater and dishwasher	12 kW
electric range	8 kW
	20 kW

Each apartment is 1000 ft^2 requiring a 3-kW load (Fig. 3-6). An arbitrary 3 kW must also be included for small appliances. The combined lighting and small appliance load is

lighting	3 kW
small appliance	3 kW
	6 kW

As shown in Fig. 3-5, the demand factor is

3 kW at 100 per cent	3	kW
3 kW at 35 per cent	1.05	kW
reduced power demand	4.05	kW

The total load for each apartment, after applying the demand factors, is

space heater and dishwasher	12	kW
electric range	8	kW
lighting and small appliances	4.05	kW
total apartment feeder load	24.05	kW

With a 240-V source, each apartment feeder must be capable of supplying approximately 100.2 A (24.05 kW/240). The correct wire size for a three-wire system is given in Fig. 2-10.

Note that the wire size for the apartment feeders can be about one or two wire sizes smaller than that of the house feeder, because of the difference in ampacity.

3-2.5 Calculating the service equipment ampacity for apartments

The demand factors can also be applied when calculating the ampacity for the service entrance equipment (conductors, main disconnect switch, etc.). Assume that it is necessary to find the service equipment ampacity for the apartment described in Section 3-2.4.

The house feeder (Fig. 3-7) has a 100 per cent demand factor and requires 30 kW. This must be added to the total for the 10 apartments. Each apartment has one space heater and one dishwasher, both with 100 per cent demand factors. Thus the fixed load for each apartment is 12 kW, and the total fixed load is 120 kW for the 10 apartments. Each apartment has one range. The demand for 10 ranges is 25 kW, as shown in Fig. 3-4.

The total lighting and small appliance load is:

(10 apartments) $\times$ (3-kW lights + 3-kW appliance) = 60 kW

As shown in Fig. 3-5, the demand factor is

3 kW at 100 per cent	3	kW
57 kW at 35 per cent	19.95	kW
reduced power demand	22.95	kW

The total service entrance ampacity requirement is

house feeder	30	kW
space heater and dishwasher	120	kW
ranges	25	kW
lights and appliances	22.95	kW
total	197.95	kW

With a 240-V source, the service entrance equipment must be capable of carrying approximately 824 A (197.95 kW/240 V). Of course, this would require bus bars rather than wire conductors. Also, many utility companies will not supply single-phase power in excess of 600 A to a single service entrance. From a practical standpoint, it might be necessary use three-phase power or make a special provision with the utility company for a double service entrance.

3-2.6 Calculating ampacity for single-family residences

The typical single-family residence does not have feeders, as such, in their electrical systems. Generally, the service entrance consists of the main disconnect switch (or circuit breaker) and overcurrent devices for the various branch circuits, all located on a single panelboard. The panelboard is at a protected, but readily accessible, location. The meter is located outside, ahead of the residential panelboard.

Although no feeders are involved, the demand factors can be applied to find the ampacity of service entrance equipment (conductors, main switch, etc.) for a single-family residence. The calculations are essentially the same as for apartment buildings. However, there are two specific exceptions. *The NEC requires that the service entrance equipment be 100 A, minimum, if the dwelling requires 10 kW or more. The NEC also requires that the service entrance equipment be 200 A, minimum, if the dwelling includes electrical space heating.*

The following is an example of how the NEC demand factors can be applied to find the correct ampacity for the service entrance equipment of a single-family residence.

Assume that the dwelling is 1500 ft^2 and has a 5-kW electric clothes dryer plus a 2-kW dishwasher. Since the dwelling is 1500 ft^2, the lighting load is 4.5 kW (1500 × 3), as shown in Fig. 3-6. An arbitrary 3 kW must also be included for small appliances, plus an arbitrary 1.5 kW for a washing machine. Since only one clothes dryer is involved, the demand factor is 100 per cent, or 5 kW. Likewise, the demand factor for a dishwasher is 100 per cent, or 2 kW. Thus the fixed load is

clothes dryer	5 kW
dishwasher	2 kW
total fixed load	7 kW

The combined lighting, small appliance, and washing machine load is

lighting	4.5 kW
small appliance	3 kW
washing machine	1.5 kW
combined	9 kW

As shown in Fig. 3-5, the demand factor is

3 kW at 100 per cent	3 kW
6 kW at 35 per cent	2.1 kW
reduced power demand	5.1 kW

The total load for the dwelling, after applying the demand factors, is

fixed load	7 kW
reduced power load	5.1 kW
total load	12.1 kW

With a 120-V source, the service entrance equipment must be capable of carrying approximately 100 A (12.1 kW/120). If the source is increased to 240 V, the current can be cut in half (approximately 50 A). However, since more than 10 kW is required with either 120 V or 240 V, 100-A service entrance equipment is required, even with the 240-V source.

Also, keep in mind that a 200-A service entrance is required if the dwelling is equipped with electric space heating, even though the total load current may be below 200 A.

3-2.7 Calculating ampacity for feeder-subfeeder systems

In the feeder systems described thus far, each panelboard is supplied by a separate feeder, and each feeder is connected directly to the service entrance equipment. Such an arrangement will not be satisfactory for large buildings, industrial plants, and so on, where the feeders must run long distances. The heavy currents normally associated with feeders, combined with long runs, produce large voltage drops between the service entrance and panelboards. (The problem of voltage drop in feeders is discussed in Section 3-3.) One solution to the problem is to use larger wire sizes for the feeders. This is not economical, however, and, in some cases, not too practical. For example, if several feeders must be used, the raceway can be overcrowded with large-sized conductors.

A more practical solution is to install a switchboard at a convenient point between the service entrance and the panelboards. Such an arrangement is shown in Fig. 3-8. One large feeder supplies the switchboard from the service entrance. Each of the panelboards is served by a subfeeder from the switchboard. The feeder current is the simple sum of the subfeeder currents. The switchboard must provide overcurrent protection for each of the subfeeders. The ampacity of the overcurrent devices must be equal to the ampacity of the corresponding subfeeder conductor.

Since voltage-drop considerations are the primary reason for a feeder–subfeeder system, design examples are given in Section 3-3.

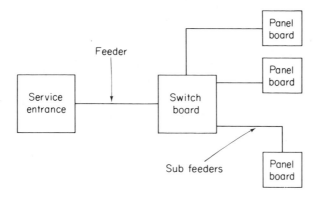

FIGURE 3-8   Wiring with switchboard and subfeeders.

3-3 CALCULATING FEEDER VOLTAGE DROP

Thus far in this chapter we have discussed feeder ampacity requirements based on anticipated loads. Although feeder conductors must be capable of carrying the full load current at all times, they must also produce a voltage drop *less than* the permitted limits. As discussed in Chapter 2, *the NEC permits a maximum 3 per cent voltage drop to the farthest outlet in a branch circuit, and a 5 per cent voltage drop if feeders are included.*

This can be interpreted to mean that feeders can have a 2 per cent drop. Of course, if the branch circuits have a 1 per cent drop, the feeders could have a 4 per cent drop, and the system would still meet NEC requirements. Except in very unusual circumstances, such an arrangement (4 per cent drop in the feeders) would be a poor design. Good design dictates that the feeder voltage drop be limited to 2 percent *maximum*. This allows a full 3 per cent voltage drop for the branch circuits.

3-3.1 Voltage drop versus source voltage

Since voltage drop is calculated as a percentage of the source voltage, a larger source voltage will permit a larger voltage drop and will still be within tolerance. For example, with a 120-V source and a 2 per cent maximum, the voltage drop could be 2.4 V in the feeder. If 240 V is used, the drop could be increased to 4.8 V, without exceeding the maximum. In practical terms, this means that a smaller wire size can be used for the feeders, with all other factors (distance, load, etc.) remaining equal. Also, a larger source voltage requires a smaller current (for a given load). This, in turn, permits a smaller wire size for the feeders.

A simple solution to the voltage-drop problem is to increase the source voltage. Although simple, the solution is not necessarily practical. In some cases, the source voltage cannot be changed. In other cases, the load must be operated from a given voltage. Both these problems can be overcome by means of transformers. However, the cost and inconvenience of transforms may outweigh the advantages of a higher voltage. In such cases, the designer must decide on a tradeoff.

To provide a background for making such a decision, we shall consider a feeder voltage-drop problem and discuss four possible solutions: operating at 115/230 V, operating at 230-V three-phase, operating at 460 V, and operating at 115/230 V with feeders and subfeeders.

3-3.2 Basic feeder voltage-drop problems

Assume that the load is 30 kW and that it can be divided into three equal (10 kW) loads if necessary. The distance from the service entrance to the loads is 500 ft. If necessary a switchboard can be installed at a convenient point between the service entrance and the load. The voltage drop must not exceed 2 per cent. The load is resistive and has a power factor of 1. Aluminum THW conductors are to be used throughout. Ignore the effects (derating) of high temperature and skin effect. However, if subfeeders are used, it is necessary to consider the derating effect of having more than three conductors in a raceway.

Voltage drop operating at 115/230 V. (See Fig. 3-9.) The first step in any solution is to convert the power into a required current.

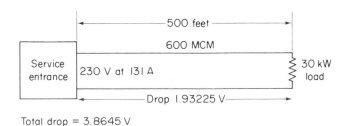

Total drop = 3.8645 V

FIGURE 3-9   **Conductor size for 30 kW load at 115-230V, single-phase.**

The 230-V source will require half the current of a 115-V source. Given the large power (30 kW) involved, the 230-V source should be used. Since the load is resistive, the power factor can be ignored. With 230-V and 30 kW, the required current is 30 kW/230 V = 131 A.

To keep the voltage drop below 2 per cent, select an arbitrary voltage drop below 2 per cent. For example, assume a voltage drop of 1.7 per cent, or 230 V × 1.7 per cent = 3.9 V.

Using the information of Section 2-7.1, find the wire size for 1000 ft (2 times the distance of 500 ft) of aluminum conductors (assumed resistivity of 17).

$$\text{cmil} = \frac{17 \times 131 \times (500 \times 2)}{230 \times 0.017} = \text{approximately 571 MCM}$$

The next largest standard size is 600 MCM.

Check the selected standard size for correct ampacity. As shown in Fig. 2-10, 600-MCM aluminum THW conductors will carry more than 131 A.

Next, find the dc resistance of the conductor, and determine the actual voltage drop. As shown in Fig. 2-6, the dc resistance for 1000 ft of 600-MCM aluminum conductors is 0.0295 Ω. With this resistance and a current of 131 A, the voltage drop is 3.8645 V (below the desired 3.9 V).

Voltage drop operating at 230 V three-phase. (See Fig. 3-10.) With the same load operated at three phase, the load must be divided into three equal parts of 10 kW each (to assure a balanced load). However, the total load remains at 30 kW, the loads are resistive, and the power factor can be ignored. With 230 V and 30 kW, the required three-phase current is 30 kW/230 V × 1.732 = 76 A.

$$\text{cmil} = \frac{17 \times 76 \times (500 \times 1.732)}{230 \times 0.017} = \text{approximately 282 MCM}$$

The next largest standard size is 300 MCM

Check the selected standard size for correct ampacity. As shown in Fig. 2-10, 300-MCM aluminum THW conductors will carry more than 76 A.

Next, find the dc resistance of the conductor, and determine the actual voltage drop. As shown in Fig. 2-6, the dc resistance for 1000 ft of 300-MCM aluminum conductors is 0.059 Ω.

The resistance of each three-phase conductor (500 ft) is 0.059 × 0.5 = 0.0295. With this resistance and a current of 76 A, the voltage drop is 2.242 V (well below the desired 3.9 V).

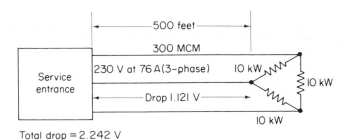

Total drop = 2.242 V

FIGURE 3-10   Conductor size for 30 kW load at 230V, 3-phase.

From these figures it is obvious that a three-phase 230-V system requires a smaller wire size, and makes it easier to meet a given voltage drop, than a single-phase 230-V system. However, three-phase is not always available, nor is it possible to operate all equipment with three-phase power.

Voltage drop operating at 460 V. (See Fig. 3-11). With 460 V and 30 kW, the required current is 30 kW/460 V = 66 A.

Assume a voltage drop of 1.7 per cent, or 460 × 1.7 per cent = 7.8 V.

$$\text{cmil} = \frac{17 \times 66 \times (500 \times 2)}{460 \times 0.017} = \text{approximately 144 MCM}$$

The next largest standard size is No. 000 AWG (Fig. 2-6).

Check the selected standard size for correct ampacity. As shown in Fig. 2-10, No. 000 AWG aluminum THW conductors will carry more than 66 A.

As shown in Fig. 2-6, the dc resistance for 1000 ft of No. 000 AWG aluminum conductors is 0.105 Ω. With this resistance and a current of 66 A, the voltage drop is 6.93 V (below the desired 7.8 V).

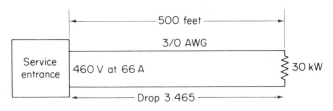

Total drop 6.93 V

FIGURE 3-11   Conductor size for 30 kW at 460V, single-phase.

Voltage drop operating at 115/230 V with subfeeders. (See Fig. 3-12.) Now assume that the 30-kW load is, in practical terms, three loads at different physical locations. Also assume that the loads are not even (one is 12 kW, one 8 kW, and one 10 kW). It is possible to install a switchboard at a common point, as shown in Fig. 3-12. The switchboard is 250 ft from the service entrance, 100 ft from the 12-kW load 250 ft from the 8-kW load, and 200 ft from the 10-kW load. The load is resistive, so the power factor can be ignored. The load must be operated from 115/230 V. No more than three conductors are required in any raceway. The maximum 2 per cent voltage drop (230 X 0.02 = 4.6 V) should be divided between the feeders and subfeeders. To simplify calculations, allow a 2-V drop for the feeders and a 2-V drop for the subfeeders.

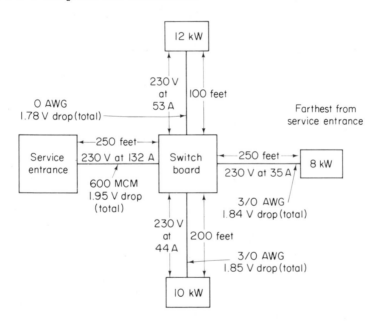

FIGURE 3-12   Conductor size for 30 kW at 230V with feeder, switchboard, and subfeeders.

Design starts with the subfeeders. The required currents for the subfeeders are

$$12 \text{ kW}/230 \text{ V} = 53 \text{ A}$$
$$8 \text{ kW}/230\text{V} = 35 \text{ A}$$
$$10 \text{ kW}/230 \text{ V} = 44 \text{ A}$$

The current for the feeder is the simple sum of the subfeeder currents, or 132 A.

The wire size for the 12-kW subfeeder is

$$\text{cmil} = \frac{17 \times 53 \times (100 \times 2)}{2} = \text{approximately } 90{,}100$$

The next largest standard size is No. 0 AWG (Fig. 2-6).

As shown in Fig. 2-10, No. 0 AWG aluminum THW conductors will carry more than 53 A. As shown in Fig. 2-6, the dc resistance for 1000 ft of No. 0 AWG aluminum conductor is 0.168 Ω. The resistance for 200 (100 $\times$ 2) feet is 0.168 $\times$ 0.2 = 0.0336 Ω. With this resistance, and a current of 53 A, the voltage drop is 1.78 V (below 2 V).

The wire size for the 8-kW subfeeder is

$$\text{cmil} = \frac{17 \times 35 \times (250 \times 2)}{2} = \text{approximately } 148{,}750$$

The next largest standard size is No. 000 AWG (Fig. 2-6).

As shown in Fig. 2-10, No. 000 AWG aluminum THW conductors will carry more than 35 A. As shown in Fig. 2-6, the dc resistance for 1000 ft of No. 000 AWG aluminum conductor is 0.105 Ω. The resistance for 500 (250 $\times$ 2) ft is 0.105 $\times$ 0.5 = 0.0525 Ω. With this resistance and a current of 35 A, the voltage drop is 1.84 V (below 2 V).

The wire size for the 10-kW subfeeder is

$$\text{cmil} = \frac{17 \times 44 \times (200 \times 2)}{2} = \text{approximately } 149{,}600$$

The next largest standard size is No. 000 AWG (Fig. 2-6).

As shown in Fig. 2-10, No. 000 AWG aluminum THW conductors will carry more than 44 A. As shown in Fig. 2-6, the dc resistance for 1000 ft of No. 000 AWG aluminum conductor is 0.105 Ω. The resistance for 400 (200 $\times$ 2) ft is 0.105 $\times$ 0.4 = 0.042 Ω. With this resistance and a current of 44 A, the voltage drop is 1.85 V (below 2 V).

It will be seen by the figures thus far that *voltage drop*, rather than ampacity, *is the critical value in finding feeder wire size.* That is, a given wire size may be capable of carrying far more current than the anticipated load, but it will produce a voltage drop that is very close to the maximum tolerance.

With the wire size selected for the subfeeder, the next step is to select the overcurrent devices for each feeder. The ampacity of the overcurrent device should be based on the wire size and on the anticipated load. The overcurrent device cannot have an ampacity greater than the wire size but must have an ampacity equal to (or greater than) the anticipated load.

For example, the 12-kW subfeeder should have an overcurrent device at the switchboard greater than 53 A (the anticipated load) but not over 120 A (the ampacity of No. 0 AWG aluminum THW conductor, three conductors in a raceway). The 8-kW subfeeder required overcurrent protection greater than 35 A but not over 155 A. The 10-kW subfeeder required overcurrent protection greater than 44 A but not over 155 A. Three 60-A circuit breakers will provide adequate protection.

The wire size for the feeder is

$$\text{cmil} = \frac{17 \times 132 \times (250 \times 2)}{2} = \text{approximately 561 MCM}$$

The next largest standard size is 600 MCM (Fig. 2-6).

As shown in Fig. 2-10, 600-MCM aluminum THW conductors will carry more than 132 A. As shown in Fig. 2-6, the dc resistance for 1000 ft of 600-MCM aluminum conductor is 0.0295 Ω. The resistance for 500 (250 $\times$ 2) ft is 0.0295 $\times$ 0.5 = 0.01475 Ω. With this resistance and a current of 132 A, the voltage drop is 1.95 V (below 2 V).

Note that the 8-kW load is farthest from the service entrance (500 ft, 250-ft feeder plus 250-ft subfeeder). The total voltage drop from service entrance to the farthest load is 1.95 (feeder) plus 1.84 (subfeeder), or 3.79 V. This is approximately 1.6 per cent (3.79/230) of the 230-V source. For reference, the voltage drops for the 12-kW and 10-kW loads are 3.73 V and 3.8 V, respectively. Both are well below the maximum 2 per cent.

3-4 BRANCH CIRCUIT DESIGN

In practical work, the branch circuits are usually designed first. Then the feeders are designed to match the branch circuits. In some ways branch circuits are easier to design than feeders. This is because the NEC has a number of specific requirements for branch circuits. In other ways, branch circuits are more difficult to design. The main problem in designing branch circuits is dividing the load.

Obviously, any one branch circuit cannot be overloaded. In addition to being poor design, this is a violation of the NEC. On the other hand, it is not economical to have more branch circuits than necessary. It is possible to get this balance in branch circuit design if several guidelines or rules of thumb are followed. Before going into these guidelines, let us examine the basic NEC requirements for branch circuits.

3-4.1 Basic NEC branch circuit requirements

Figure 3-13 shows the basic NEC requirements for branch circuits. As shown, there are five ampacity ratings: 15, 20, 30, 40 and 50 A. These ratings represent the *maximum* to be carried by the branch circuit, as well as the required overcurrent protection for each branch circuit. For example, a 15-A branch circuit requires 15-A overcurrent protection and must be used only where the maximum anticipated current is 15 A. If the maximum current exceeds 15 A, say 18 A, a 20-A branch circuit must be used. If the maximum current exceeds 50 A, at least two branch circuits must be used.

These maximum ratings must be derated to 80 per cent of their ampacity if the load is continuous. For example, a 15-A branch circuit is required for a continuous load of 12 A, a 20-A branch circuit is required for continuous loads of 16 A, and so on.

The wire sizes shown in Fig. 3-13 are for copper. If aluminum is

Device	Branch circuit rating				
Minimum size (AWG) for copper conductors:	15 A	20 A	30 A	40 A	50 A
Circuit wires	14	12	10	8	6
Taps	14	14	14	12	12
Fixture wires	18	18	14	12	12
Outlet devices:					
Receptacle rating	15 A	15 or 20 A	30 A	40 or 50 A	50 A
Lampholders	Any type	Any type	Heavy duty	Heavy duty	Heavy duty
Maximum load and overcurrent protection	15 A	20 A	30 A	40 A	50 A
Continuous load	12 A	16 A	24 A	32 A	40 A

FIGURE 3-13   NEC branch circuit ampacity ratings.

used, larger wire sizes are required to maintain the ampacity. For example, in a 15-A branch circuit, using aluminum conductors, the circuit wires and taps must be at least No. 12 AWG.

Keep in mind that these wire sizes are minimum. It may be necessary to increase the wire size, in special cases, to keep the voltage drop within tolerance. Also, remember that a tap wire can never be larger than a circuit wire. Furthermore, if it is necessary to increase the circuit wire size, the taps and fixture wires must be increased accordingly. Otherwise, the smaller fixture and tap wires would not have adequate protection. Wires used for permanent fixtures can never be larger than a circuit or a tap wire.

The ampacity of outlet receptacles must match the ampacity of the branch circuit. There are two exceptions to this rule. In some cases 15-A receptacles can be used in 20-A branch circuits; 50-A receptacles can be used in 40-A branch circuits.

Any type of lampholder can be used on 15- and 20-A branch circuits. However, only heavy-duty lampholders can be used on 30-, 40-, and 50-A branch circuits. Heavy-duty lampholders are defined as those with the heavy-duty, or mogul, sockets. Such heavy-duty sockets are used on lamps with power ratings over 300 W (300 to 1500 W). Lamps with power ratings below 300 W used medium-base lampholders.

3-4.2 40- and 50-A branch circuits

In dwellings (both apartments and single-family units), the 40- and 50-A branch circuits are generally used *only for electric ranges*, and possibly for infrared heating units. As noted earlier, a typical electric range could draw up to about 12 kW. Assuming a 230-V supply, the current is about 50 A.

In applications *other than dwellings*, 40- and 50-A branch circuits can also be used for heavy-duty lampholders or for any use that requires large currents. However, if medium-base lamps (below 300 W) must be used in any application, they cannot be tapped into 40- or 50-A branch circuits.

3-4.3 30-A branch circuits

In dwellings, the 30-A branch circuits are generally limited to those portable appliances that require higher currents than are avail-

able on 15- and 20-A branch circuits.   Such appliances, typically
6-kW clothes dryers, cannot be rated over 24 A (which is 80 per cent
of 30A), since they are considered to be a continuous load.   The plugs
for such appliances must be rated at 30 A to match the branch circuit receptacles.   This prevents these portable appliances from being
connected (accidently) to lower (15-A, 20-A) or higher (40-A, 50-A)
branch circuits.

The matching of receptacles is a good design practice.   If a 24-A
appliance is connected to a 15- or 20-A branch circuit, the overcurrent
device will be actuated.   If the 24-A appliance is connected to a 40-
or 50-A branch circuit, a fault in the appliance could destroy the
appliance or cause a fire without opening the higher-rated overcurrent
device.

In applications other than dwellings, 30-A branch circuits can
also be used for heavy-duty lampholders and for any use that requires
large currents (30 A maximum or 24 A continuous), except lampholders with medium-size sockets.

3-4.4 15- and 20-A branch circuits

In dwellings, the 15- and 20-A branch circuits are used extensively
for lighting and small appliances.   In those dwellings without electric
ranges, electric space heating, and certain clothes dryers, 15- and 20-A
branch circuits are used exclusively.

Any size of lampholder can be used on 15- and 20-A branch circuits, since the overcurrent limit is low enough to protect any one
lighting unit.   For example, assume that a mogul-base lampholder is
tapped to a 15-A branch circuit and that a 1500-W lamp is used with
a source of 115 V (all these are extreme cases).   The resultant current
flow would be about 13 A.   Even if the load is left on continuously,
the load will not trip the overcurrent.

In theory there is no maximum number of 15- and 20-A branch
circuits that can be used.   In practical terms, however, the number of
overcurrent devices on any one panelboard is limited to 42.   However, *the NEC requires one 20-A branch circuit (each) for the kitchen,
dining room, pantry, and laundry room area.*   The 20-A branch for the
laundry can be omitted in apartments where there is a common
laundry area (and each apartment does not have its own washer and
dryer).   *The NEC also requires at least one branch circuit for lighting
only, for each 500 ft^2 of area.*

A 15-A branch circuit must have 15-A receptacles. A 20-A branch circuit can have either 15- or 20-A receptacles. This simplifies the use of small portable appliances.

3-4.5 Connecting appliances to branch circuits

If a permanently installed appliance is connected to a branch circuit and the appliance is the only load on that branch circuit, the current rating of the appliance can not exceed 80 per cent of the branch circuit rating. For example, if a 15-A branch circuit is used exclusively for a permanently installed appliance, the appliance can have a current rating up to 12 A.

If a permanently installed appliance is used on a branch circuit with other loads, the current rating of the appliance cannot exceed 50 per cent of the branch circuit. Using the same 15-A branch circuit example, the appliance current rating must be 7.5 A or less.

If a portable appliance is used, the current rating of the appliance should not exceed 80 per cent of the branch circuit. Most portable appliances are designed to ensure this condition. That is, an appliance with a 15-A plug will not draw more than 12 A. Of course, the designer has no control over the user of portable appliances. There is no way of preventing two 12-A appliances from being connected at two receptacles of the same 15-A branch circuit. However, the overcurrent device will be tripped, and there will be no danger to the appliances, or wiring.

3-5 BRANCH CIRCUIT LOADS AND VOLTAGE DROPS

As discussed, the NEC permits a 3 per cent voltage drop to the farthest outlet in a branch circuit. This drop is measured from the panelboard to the outlet. An alternative method for computing maximum voltage drop is 5 per cent from the farthest outlet to the service entrance. This voltage drop includes the feeder. In theory, if the feeders produced a 1 per cent drop, the branch circuit could be 4 per cent. However, good design dictates that the branch circuit voltage drop (panelboard to farthest outlet) be not more than 3 per cent.

Figure 3-13 shows the NEC requirements for minimum wire size in branch circuits. All branch circuits should be designed with the assumption that the wire sizes given in Fig. 3-13 will be used. That is, the length of branch circuits should be limited so that the voltage drop, using the Fig. 3-13 wire sizes, will be less than 3 per cent.

If the branch circuit has only one tap (receptacle, permanently wired fixture, etc.), the voltage drop calculations are fairly simple (somewhat like those for feeders, Section 3-3).

If the branch circuit has several taps, the voltage drop calculations become quite complex. For example, as shown in Fig. 3-14, a 15-A branch circuit has three outlets, each with a 3-A load. The current in the conductors is as follows: 9 A between the panelboard and the first outlet, 6 A between the first and second outlets, and 3 A between the second and third outlets. Assume that No. 14 AWG copper wire is used (Fig. 3-13), that the resistance is 2.57 Ω per 1000 ft, and that the distances are as shown in Fig. 3-14 (150 ft total).

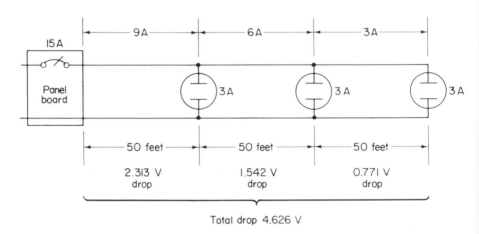

FIGURE 3-14   Example of voltage drops in branch circuits.

The voltage drop between panelboard and the first outlet is 100/1000 = 0.1; 0.1 $\times$ 2.57 = 0.257 Ω; 0.257 $\times$ 9 = 2.313 V.

The voltage drop between the first and second taps is 0.257 $\times$ 6 = 1.542 V. The voltage drop between the second and third taps is 0.257 $\times$ 3 = 0.771 V. The total voltage drop is 4.626 (2.313 + 1.542 + 0.771). If the source is 115 V, the maximum permitted voltage drop is 3.45 V. Thus the circuit of Fig. 3-14 would not meet NEC requirements if the source is 115 V and No. 14 AWG copper wire is used. If the source cannot be raised to 230 V, the wire size must be increased.

This example is oversimplified. The voltage-drop calculations become very complex when the loads, and the distances between taps, are unequal.

3-5.1 Branch circuit guidelines for voltage drop

There are some rules of thumb or guidelines that can be applied when designing branch circuits. These rules will keep the voltage drop within tolerance and still permit the use of wire sizes shown in Fig. 3-13. The following is a summary of these guidelines.

Keep the number of taps on any branch circuit at 6 (or less) and never more than 8.

With a 115-V source, keep the branch circuit length to the final outlet at

1. Less than 40 ft if the final outlet carries more than 80 per cent of the maximum-rated load.
2. Less than 60 ft if the final outlet carries between 50 and 80 per cent of the maximum-rated load.
3. Less than 90 ft if the final outlet carries up to 50 per cent of the maximum-rated load.

If the source voltage is increased, the length may be increased in proportion. For example, if the source voltage is raised to 230 V from 115 V, the lengths may be doubled to 80, 120, and 180 ft.

With any source voltage, keep the branch circuit length from the panelboard to the first outlet at one third (or less) of the total length of the branch circuit.

As an example, assume that the source voltage is 115 V and that the final outlet carries between 50 and 80 per cent of the total load. Under these conditions, the branch circuit should be no longer than 60 ft total, from panelboard to the final outlet, and no more than 20 ft from panelboard to the first outlet.

Keep in mind that these are guidelines. The voltage drop should be verified by calculation, especially if there are unusual circumstances (such as long runs between outlets, from panelboard to the first outlet, etc.).

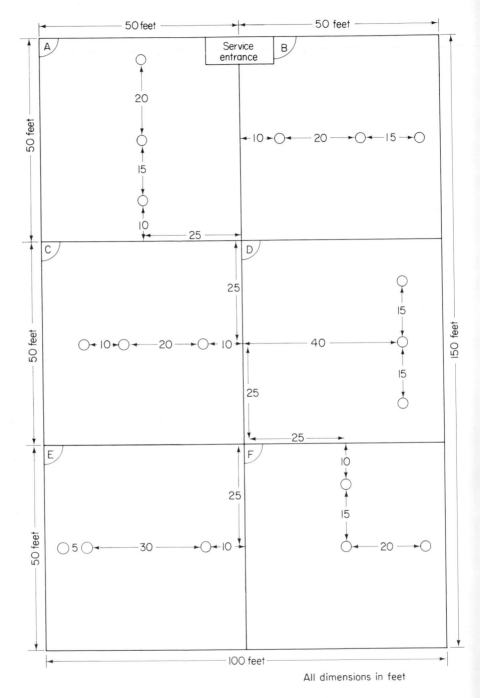

All dimensions in feet

FIGURE 3-15 Physical location of service entrance and loads in storage area.

3-6 EXAMPLE OF BRANCH CIRCUIT, FEEDER, AND OVERCURRENT DESIGN

Assume that a storage area similar to that shown in Fig. 3–15 is to be provided with electrical power. The total area is 15,000 ft^2 (150 × 100 ft). The storage area is divided into six equal compartments (A, B, C, D, E, and F). Each compartment is 2500 ft^2 (50 × 50 ft) and must be provided with a separate branch circuit. Each branch circuit has three receptacles, located as shown, and each receptacle must be capable of delivering a *continuous* 5 A of current. A panelboard can be located anywhere in the storage area, but the service entrance must be at the end of the building, as shown. The power source is 115/230 V. Copper TW conductors are to be used throughout.

Find a practical location for the panelboard and routing of the branch circuit raceways. Calculate the required branch circuit ratings. Show the raceway routing for the feeders between the service entrance and panelboard. Finally, calculate the feeder wire size and overcurrent protection for the feeders.

3-6.1 Branch circuit ratings

Each branch circuit must be capable of delivering a *continuous* 15 A of current (three receptacles at 5 A each). A 20-A branch circuit rating is thus required in each of the six compartments. (A 20-A branch circuit will provide a continuous 16 A, or 80 per cent of 20-A.)

3-6.2 Branch circuit overcurrent protection

Since each of the six branch circuits is 20 A, the panelboard must be provided with six 20-A circuit breakers.

3-6.3 Branch circuit length

Since all receptacles or outlets provide the same current (5 A), the final outlet carries up to 50 per cent of the maximum rated load. Using the guidelines of Section 3-5.1, each branch circuit should not exceed 90 feet if 115 V is used. In our example, we are using 230 V, so the 90-ft length can be doubled to 180 ft for the final outlet. The first outlet must be no more than one-third of the total branch circuit length, or 60 ft (180/3 = 60).

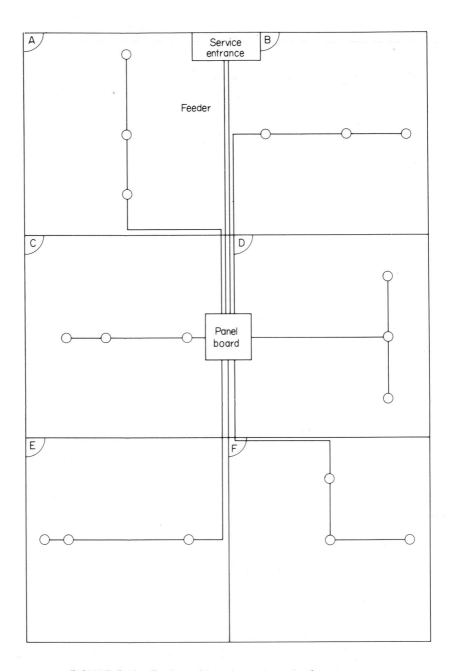

FIGURE 3-16   Feeder and branch circuit routing for storage area.

By placing the panelboard in the center of the building, as shown in Fig. 3-16, it is possible to keep the length of each branch circuit to the farthest receptacle or outlet at less than 95 ft and the distance to the first outlet at no more than 60 ft.

In compartment A the length is 25 ft from panelboard to the compartment corner, then 25 ft to the centerline of the compartment, 10 ft to the first receptacle, 15 ft to the second receptacle, and 20 ft to the final receptacle. The same is true of compartments B, E, and F. The distance from the panelboard to the final outlet in compartments C and D is even less (40 and 55 ft, respectively).

Note that all raceways are bent at 90-degree angles and that no raceway requires a 360-degree bend (from panelboard to farthest outlet).

3-6.4 Branch circuit wire sizes

As shown in Fig. 3-13, the circuit wires for a 20-A branch circuit must be No. 12 AWG (when copper is used). Assume that No. 12 AWG copper TW is used from the panelboard directly to each of the receptacles.

3-6.5 Branch circuit voltage drops

Although the guidelines of Section 3-5.1 are calculated so that the branch circuit voltage drops will be less than 3 per cent (from panelboard to farthest outlet), let us verify this by calculating the voltage drop for branch circuits in each of the compartments. Note that none of the branch circuits are longer than that for compartment A. However, it is possible that the voltage drop will be greater in one or more of the other compartments.

Compartment A. The resistance for No. 12 AWG copper wire is 1.62 Ω/1000 ft. The distance from the panelboard to the first outlet in compartment A is 60 ft, which necessitates the use of 120 ft of conductor. The voltage drop between the panelboard and the first outlet is 120/1000 = 0.12; 0.12 $\times$ 1.62 = 0.1944; 0.194 $\times$ 15 A = 2.916 V.

The distance from the first outlet to the second outlet is 15 ft, so 30 ft of conductor is required. The voltage drop between the first and second outlets is 30/1000 = 0.03; 0.03 $\times$ 1.62 = 0.0486 Ω; 0.0486 $\times$ 10 A = 0.486 V.

The distance between the second and third outlets is 20 ft, so 40 ft of conductor is required. The voltage drop between the second and third outlets is 40/1000 = 0.04; 0.04 × 1.62 = 0.0648 Ω; 0.0648 × 5 A = 0.324 V.

The total compartment A voltage drop is 3.726 V (2.916 + 0.486 + 0.324). With a source of 230 V, the maximum permitted voltage drop is 6.9 V.

Compartment B. The distance from the panelboard to the first outlet in compartment B is 60 ft, and the voltage drop is 2.916 V, as for compartment A.

The distance from the first outlet to the second outlet is 20 ft, so 40 ft of conductor is required. The voltage drop between the first and second outlets is 40/1000 = 0.04; 0.04 × 1.62 = 0.0648 Ω; 0.0648 × 10 A = 0.648 V.

The distance between the second and third outlets is 15 ft, so 30 ft of conductor is required. The voltage drop between the second and third outlets is 30/1000 = 0.03; 0.03 × 1.62 = 0.0486 Ω; 0.0486 × 5 A = 0.243 V.

The total compartment B voltage drop is 3.807 V (2.916 + 0.648 + 0.243). Although the total distance is the same, the voltage drop for compartment B is greater than that for compartment A (but well below the maximum 6.9 V).

Compartment C. The distance from the panelboard to the first outlet in compartment C is 10 ft, so 20 ft of conductor is required. The voltage drop between the panelboard and the first outlet is 10/1000 = 0.01; 0.01 = 1.62 = 0.0162 Ω; 0.0162 × 15 A = 0.243 V.

The distance from the first outlet to the second outlet is 20 ft, so 40 ft of conductor is required. The voltage drop between the first and second outlets is 40/1000 = 0.04; 0.04 × 1.62 = 0.0648 Ω; 0.0648 × 10 A = 0.648 V.

The distance between the second and third outlets is 10 ft, so 20 ft of conductor is required. The voltage drop between the second and third outlets is 20/1000 = 0.02; 0.02 × 1.62 = 0.324 Ω; 0.324 × 5 A = 0.162 V.

The total compartment C voltage drop is 1.053 (0.243 + 0.648 + 0.162).

Compartment D. The distance from the panelboard to the first outlet in compartment D is 40 ft, so 80 ft of conductor is required. The voltage drop between the panelboard and the first outlet is 80/1000 = 0.080; 0.080 × 1.62 = 0.1296 Ω; 0.1296 × 15 A = 1.944 V.

The distance from the first outlet to either of the remaining outlets is 15 ft, so 30 ft of conductor (to each of the other outlets) is required. The voltage drop from the first outlet to either of the other outlets is 30/1000 = 0.03; 0.03 × 1.62 = 0.0486 Ω; 0.0486 × 5 A = 0.243 V.

Compartment E. The distance from the panelboard to the first outlet in compartment E is 60 ft, and the voltage drop is 2.916, as for compartment A.

The distance from the first outlet to the second outlet is 30 ft, requiring 60 ft of conductor. The voltage drop between the first and second outlets is 60/1000 = 0.06; 0.06 × 1.62 = 0.0972; 0.0972 × 10 A = 0.972 V.

The distance between the second and third outlets is 5 ft, requiring 10 ft of conductor. The voltage drop between the second and third outlets is 10/1000 = 0.01; 0.01 × 1.62 = 0.0162 Ω; 0.0162 × 5 A = 0.081 V.

The total compartment E voltage drop is 3.969 V (2.916 + 0.972 + 0.081). Note that compartment E has the highest voltage drop thus far.

Compartment F. The distances for compartment F are the same as for compartment A, although the physical arrangement is different. Thus the total voltage drop for compartment F is 3.726 V.

3-6.6 Feeder wire sizes

The feeders run 75 ft from the service entrance (at the end of the building) to the panelboard (at the building's center). The total continuous load is 90 A (15 A for each of the six compartments).

Using the information of Section 2-7.1 and assuming a feeder voltage drop of 1.7 per cent (230 V × 1.7 = 3.9 V), the wire size for 150 ft (2 times the distance of 75 ft) of copper conductors (with assumed resistivity of 10.4) is

$$\text{cmil} = \frac{10.4 \times 90 \times (75 \times 2)}{320 \times 0.017} = 36{,}000$$

The next largest standard size is No. 4 AWG (Fig. 2-6).

Check the selected standard size for correct ampacity. As shown in Fig. 2-10, No. 4 AWG copper conductors will not carry 90 A. The next largest wire size (of copper TW) that will carry 90 A is No. 2 AWG.

Next, find the dc resistance of the conductors and determine the actual voltage drop. As shown in Fig. 2-6, the dc resistance for 1000 ft of No. 2 AWG is 0.162 Ω. The resistance for 150 ft of No. 2 AWG is 0.15 $\times$ 0.162 = 0.0243 Ω. With this resistance and a current of 90 A, the voltage drop is 2.187 V (below the maximum 3.9 V).

3-6.7 Total voltage drop

With a feeder voltage drop of 2.187 V and a branch circuit voltage drop (to the farthest outlet in compartment E) of 3.969 V, the total voltage drop is 6.156 V. This is well below the NEC maximum of 5 per cent (230 V $\times$ 5 = 11.5 V).

3-6.8 Feeder overcurrent protection

The ampacity of the feeder overcurrent device is based on wire size (No. 2 AWG) and type (TW), as well as the anticipated load (90 A). The overcurrent device cannot have an ampacity greater than the wire size but must have an ampacity equal to (or greater than) the anticipated load.

In our example the feeder should have an overcurrent device greater than 90 A (the anticipated load) but not over 95 A (the ampacity of No. 2 AWG copper TW conductor, three conductors in a raceway). A 95-A circuit breaker should provide adequate protection.

Chapter 4

Grounding Problems
in Electrical Wiring

Most electrical systems and equipment require some form of *grounding*. Although the term "ground" implies an electrical connection to earth, the terms *common connection* or *common ground* might be more accurate. Figures 4-1 and 4-2 illustrate the difference.

Figure 4-1 shows the power distribution system typical of those found in electronic devices. The same system could be applied to any installation where there is a *common metal frame* that acts as one conductor.

In the system of Fig. 4-1, the positive terminal of the power

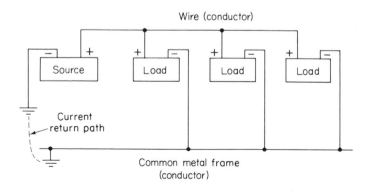

FIGURE 4-1 Power distribution system typical for electronic devices, automobiles, etc., (not suitable for wiring per NEC requirements).

source is connected to all the loads by means of a wire or wires. The negative terminal of the source is connected directly to the common metal frame, such as the chassis of an electronic device, thus making the system of Fig. 4-1 a *negative ground installation*. In turn, the chassis or common metal frame may or may not be connected to true earth ground. However, there is current flow since there is a path for the current to leave the source and return to the source. Sometimes, the negative path of Figure 4-1 is called the *ground return*. The system can be adapted to a positive ground installation by reversing the polarity of the supply and all the loads.

The circuit of Fig. 4-1 shows that an electrical system will function with or without a true earth ground and with only one wire as a conductor, as long as there is a return path provided by a common conductor (or common ground). However, the system shown in Fig. 4-1 is not acceptable to the NEC. *The NEC requires that insulated con-*

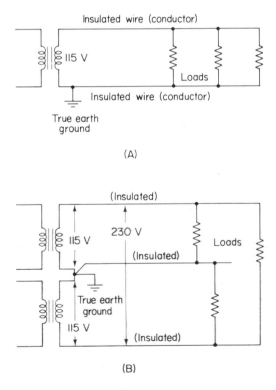

FIGURE 4-2   Basic grounding systems for wiring that meets NEC requirements.

ductors must be used to carry current to a load and from the load back to the power source. Since it is not practical to insulate the metal chassis of electronic devices, metal automobile frames, and any other large metal surface, the circuit of Fig. 4-1 cannot be used for electrical wiring covered by NEC regulations.

The circuits of Fig. 4-2 show electrical grounding systems that do meet basic NEC requirements. (The specific details for grounding are discussed throughout this chapter.)

In Fig. 4-2a the power source is a transformer secondary winding that produces 115 V. Note that one terminal of the transformer is at true earth ground. Thus all the loads connected to this terminal (through insulated conductors) are at true earth ground. That is, *one terminal* of each load is at true earth ground.

In Fig. 4-2b the power source is at 230 V and is provided by two 115-V transformer windings connected in series. Note that the common point between the two windings is connected to true earth ground. Thus any load terminals connected to the center or neutral point are at true earth ground.

With either circuit of Fig. 4-2, a *system ground* is established, since one terminal of the power supply is connected through a low-resistance path to true earth ground.

4-1 BASIC NEC SYSTEM GROUNDING REQUIREMENTS

Contrary to popular opinion, not all electrical systems covered by NEC must be grounded. In fact, the NEC is somewhat fuzzy in this regard. For example, the NEC will permit some high-voltage systems to be ungrounded (as will some, but not all, local authorities).

The NEC does say that a system must be grounded if, after grounding, the "voltage to ground" is less than 150 V, and recommends that the system be grounded if the voltage to ground can be less than 300 V.

The NEC also states that whatever grounding system is used, the voltage to ground must be the minimum possible voltage. Further, the voltage to ground from any ungrounded point must be the same as the voltage to any grounded conductor.

As we shall see, there are simple methods to meet these basic grounding requirements.

4-1.1 Voltage-to-ground problem

Before going on, let us define the voltage-to-ground problem. A ground can be true earth ground or any conductor connected directly or indirectly to earth ground. For example, metal pipes (water, gas, drain, etc.) in a building are connected to other pipes, which, in turn, are buried in the ground. The metal frames of buildings (and any conductor touching the metal frames) are installed in the ground. Of course, some metal frames are sunk in concrete and do not touch true earth ground. However, when concrete is wet or even damp, it can conduct current. Thus metal pipes and metal building frames can be considered as grounds.

Note that, as a practical matter, *the NEC requires that electrical system grounds be made at metal pipes rather than at metal frames*, since buried pipes make for a sure ground. When metal pipes are not readily accessible, a ground for electrical wiring can be made by driving a metal rod into the ground or by burying metal plates in the ground and connecting the wiring to the rod or plate with a suitable conductor.

The voltage to ground is the voltage from any point in the electrical system to any object (metal pipe, metal frame, etc.) that itself is grounded. Let us see how voltage to ground is measured in some typical systems.

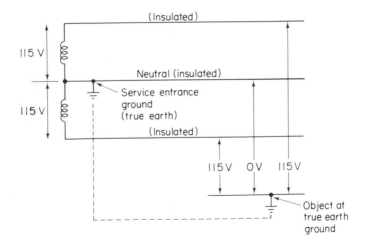

FIGURE 4-3   Single-phase, 3-wire distribution with correct NEC grounding.

4-1.2 Single-phase three-wire ground

Figure 4-3 shows the typical 115/230-V three-wire power supply generally used when a single-phase source is required. Note that the *neutral wire is grounded.* If you measure from ungrounded wire to ground or to the grounded wire, the voltage will be 115 V. Thus the voltage to ground is the minimum possible voltage, and the voltage from any ungrounded conductor is the same as the voltage from all other conductors.

If either of the other wires is grounded, two NEC requirements are violated. This is shown in Fig. 4-4, where only the bottom wire is grounded.

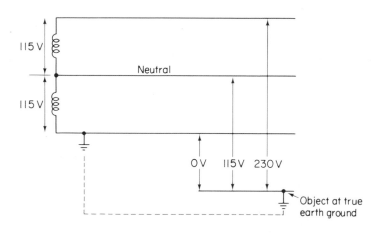

FIGURE 4-4   Single-phase, 3-wire distribution with incorrect NEC grounding (NEC violation).

If you measure from the center or neutral wire to ground, the voltage will be 115 V, which is acceptable. However, if you measure from the top wire to ground, the voltage is 230 V. Thus the voltage to ground from all ungrounded conductors is not the minimum possible, and the voltage to ground from one wire is not the same as from the other wire.

4-1.3 Three-phase four-wire ground

Figure 4-5 shows the typical 120/208-V four-wire power supply often used for three-phase systems. Note that again the neutral wire is grounded. In three-phase four-wire systems, *the NEC has an*

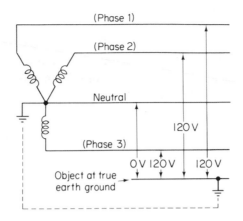

FIGURE 4-5   3-phase, 4-wire distribution with correct NEC ground.

additional requirement that each single-phase load be supplied with one grounded conductor.

If you measure from any ungrounded wire to ground, the voltage will be 115 V. Thus the voltage is the minimum possible and the same for all conductors (each load has one grounded conductor).

4-1.4 Three-phase three-wire ground

Technically, an ungrounded three-phase three-wire system does not violate the NEC requirements for grounding. Such a system is shown in Fig. 4-6. If you measure from any wire to ground, the voltage will be 0 V. Thus the voltage is the minimum possible and

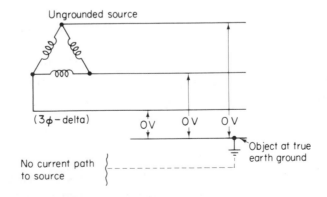

FIGURE 4-6   3-phase, 3-wire distribution without ground.

the same for all conductors. Although this does meet NEC requirements and is found in some installations, it is not necessarily the best
possible arrangement.

As stated, the NEC *recommends* that grounding be used where the
voltage to ground could be less than 300 V. For example, if there is
an accidental ground (say, because of poor insulation on one wire in
a conduit or because of excessive moisture that shorts one wire to
ground), the voltage from the other two wires to ground will be
115 V. This condition is shown in Fig. 4-7.

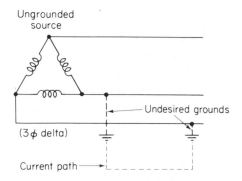

FIGURE 4-7   3-phase, 3-wire distribution with undesired grounds (providing a current path between two phases).

If a three-phase three-wire system is grounded, *any one of the
conductors* can be used for the ground. This meets the NEC requirements that the voltage to ground be minimum and the same for all
conductors.

4-1.5 System ground considerations

You may wonder why it is necessary to ground any electrical
wiring system. The answer is *safety*. System grounding of electrical
wiring provides a measure of safety should there be a *fault outside
the building*.

For example, a system ground (where one terminal of the power
supply or source is grounded) for electrical wiring acts like a lightning rod. That is why *the NEC requires that the system ground be
made at the point of service entrance* (where the electrical power
enters the building). Should the distribution or service lines be
struck by lightning, the service entrance ground will provide a ready

path to true earth ground *outside* the building, without the danger of heavy current passing through the interior wiring.

4-2 BASIC NEC EQUIPMENT GROUNDING REQUIREMENTS

Unlike the system grounding discussed in Section 4-1, *the NEC requires that all equipment in an electrical system be grounded.* That is, conduits, raceways, fittings, boxes, switch housings, fuse, and circuit breaker enclosures must be grounded. Such grounding is called *equipment grounding.*

Further, *the NEC requires that all equipment be grounded to the same earth ground as the system ground.* The only equipment exempt from grounding is equipment so isolated that no person could touch a grounded object and the electrical equipment simultaneously.

As in the case of system grounding, the main reason for equipment grounding is *safety*. In properly designed electrical wiring, there must be no contact between the electrical equipment (raceways, enclosures, etc.) and the conductors. However, there is always the possibility of an accidental ground, even with a supposedly ungrounded system. The electrical system will still operate if only one of the conductors becomes grounded. However, if a person should be in contact with a true earth ground (say standing on damp cement) and any equipment metal that has been in contact with a conductor, they will receive the full voltage.

In Fig. 4-8 the conduit has made accidental contact, or ground, with one ungrounded conductor. In turn, the conduit is in contact with the switch box and the switch plate (through the plate attaching screws). If a person in contact with true earth ground should touch the switch plate, current will flow from the fuse, through the conduit, switch box, switch plate, and the person, back to the grounded end of the transformer winding. Such a current flow is known as a *fault current*, since it only occurs when there is a fault in the system. The purpose of grounding is not to eliminate fault currents but to use the fault currents to open protective overcurrent devices (fuses, circuit breakers, etc.).

Of course, grounding all equipment in an electrical system will not prevent accidental grounding of ungrounded conductors. However, should there be an accidental ground, the fuse or other overcurrent device will open the circuit, thus eliminating the hazard.

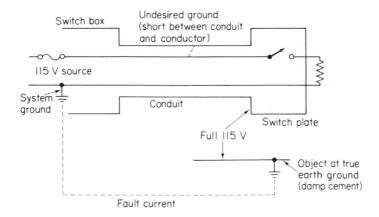

FIGURE 4-8   Effect of short between ungrounded switchplate and conductor, in distribution without equipment ground.

Such a condition is shown in Fig. 4-9.   Here the conduit is grounded to the system ground at the service entrance (as is required by NEC). The circuit breaker is in the ungrounded conductor (also required by NEC). There has been an accidental ground of the ungrounded conductor at the opposite end of the conduit (where the conduit enters the switch box).   Under these conditions, current flows through the circuit breaker, opening the circuit.

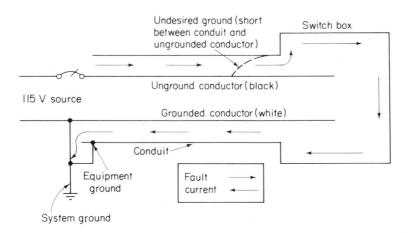

FIGURE 4-9   Effect of short between ungrounded conductor and conduit, in distribution with equipment ground.

4-3 COMBINED SYSTEM AND EQUIPMENT GROUNDS

The most practical method for grounding both the system and equipment is to provide a common ground point in the service entrance. Usually this is a terminal bonded to the service entrance enclosure, as shown in Fig. 4-10. In turn, the terminal (sometimes known as a *lay-in lug)* is connected to a true earth ground (usually a metal pipe) by a conductor of suitable size. As illustrated, the ground conductor size is determined by the size of the service conductor.

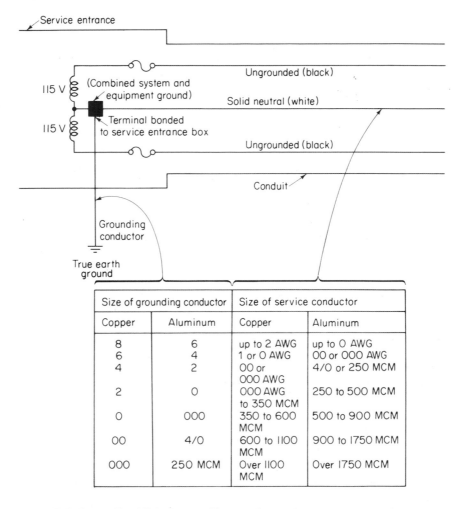

Size of grounding conductor		Size of service conductor	
Copper	Aluminum	Copper	Aluminum
8	6	up to 2 AWG	up to O AWG
6	4	1 or O AWG	OO or OOO AWG
4	2	OO or OOO AWG	4/O or 250 MCM
2	O	OOO AWG to 350 MCM	250 to 500 MCM
O	OOO	350 to 600 MCM	500 to 900 MCM
OO	4/O	600 to 1100 MCM	900 to 1750 MCM
OOO	250 MCM	Over 1100 MCM	Over 1750 MCM

FIGURE 4-10   NEC size requirements for service entrance grounding conductors.

For example, if a copper service conductor is No. 2 AWG or smaller, the required size for a copper ground conductor is No. 8 AWG.

Note that in Fig. 4-10 the neutral system conductor is grounded (as required by NEC) at the service entrance. *The NEC also requires that the neutral conductor not be grounded at any point beyond the service entrance.* If the neutral conductor is grounded at any other point, there is a possibility that some current could flow through equipment.

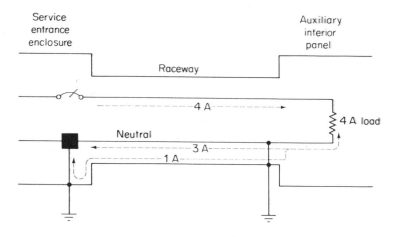

FIGURE 4-11   Effect of grounding neutral conductor at some point in distribution system after the service entrance ground.

This condition is illustrated in Fig. 4-11. As shown, the neutral conductor is grounded at the main service entrance and at an interior panel. The service enclosure and the panel are connected by a race-way. Assume that 4 A is flowing through the load and assume that the resistance of the equipment (service entrance, raceway, and aux-iliary panel) is three times that of the neutral conductor (between the two ground points). Under these conditions, the return current through the neutral conductor is 3 A, with 1 A flowing through the equipment, since the two current paths are in parallel.

4-3.1 Grounding interior equipment

When it becomes necessary to ground interior equipment, this should be done with a separate conductor, as shown in Fig. 4-12. The size of equipment grounding conductors is determined by the

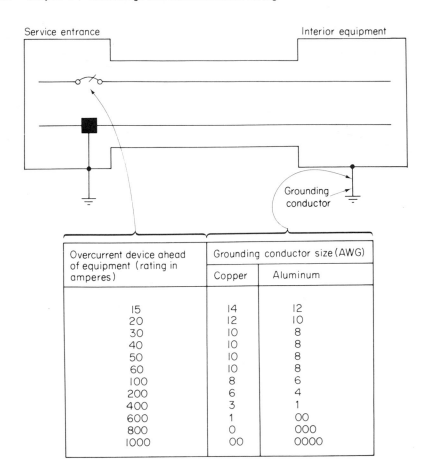

FIGURE 4-12   NEC size requirements for interior equipment grounding conductors.

Overcurrent device ahead of equipment (rating in amperes)	Grounding conductor size (AWG)	
	Copper	Aluminum
15	14	12
20	12	10
30	10	8
40	10	8
50	10	8
60	10	8
100	8	6
200	6	4
400	3	1
600	1	00
800	0	000
1000	00	0000

rating (in amperes) of the *overcurrent device ahead of the equipment* rather than by the size of the service conductors in the equipment. For example, as shown in Fig. 4-12, if a 15-A fuse or circuit breaker is used ahead of the equipment, the interior equipment grounding conductor must be No. 14 AWG (for copper) or No. 12 AWG (for aluminum).

4-3.2 Equipment grounding for ungrounded systems

As discussed in Section 4-1, all systems are not grounded. Specifically, the three-phase three-wire system does not require a ground.

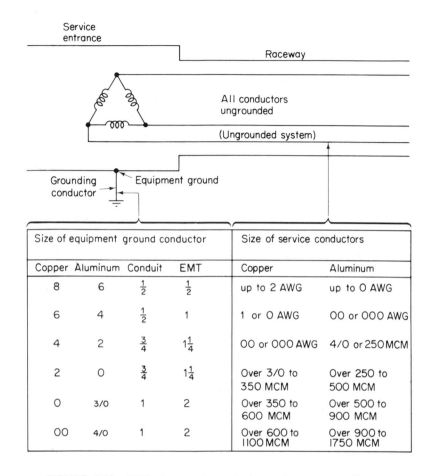

Size of equipment ground conductor				Size of service conductors	
Copper	Aluminum	Conduit	EMT	Copper	Aluminum
8	6	$\frac{1}{2}$	$\frac{1}{2}$	up to 2 AWG	up to O AWG
6	4	$\frac{1}{2}$	1	1 or O AWG	OO or OOO AWG
4	2	$\frac{3}{4}$	$1\frac{1}{4}$	OO or OOO AWG	4/O or 250 MCM
2	O	$\frac{3}{4}$	$1\frac{1}{4}$	Over 3/O to 350 MCM	Over 250 to 500 MCM
O	3/O	1	2	Over 350 to 600 MCM	Over 500 to 900 MCM
OO	4/O	1	2	Over 600 to 1100 MCM	Over 900 to 1750 MCM

FIGURE 4-13   NEC size requirements for equipment grounding conductors for ungrounded distribution systems.

However, the equipment must be grounded, even in an ungrounded system. This is shown in Fig. 4-13. With an ungrounded system, the size of the equipment grounding conductor is determined by the size of the service conductor.

4-4 PROBLEMS OF DEFECTIVE EQUIPMENT GROUNDING

If equipment components are not properly installed, there is a possibility that fault currents (due to an accidental ground) will not be sufficient to open overcurrent devices. Raceways and their associated hardware (connectors and fittings) have low resistance. Most raceways are made of steel or aluminum conduit or tubing. These materials have low resistance and, when the raceway components are *properly installed together*, the overall resistance from one end of the equipment remains low. Under these conditions, a fault current will open the overcurrent device and prevent the raceway from becoming a voltage-to-ground hazard.

If the raceway components, such as clamps, threaded fittings, connectors, bushings, and locknuts, are not making proper electrical contact even though they are mechanically secure, it is possible that the overall resistance of the equipment will increase to a point where insufficient fault currents will flow. Such a condition is shown in Fig. 4-14. Note that this circuit is similar to that of Fig. 4-10 except that there is some resistance in the raceway. Assume that this resistance is caused by the raceway not making good electrical contact with an interior panel, and that the resistance produced by this poor contact is about 5 Ω.

The condition of poor electrical contact, in itself, will not affect

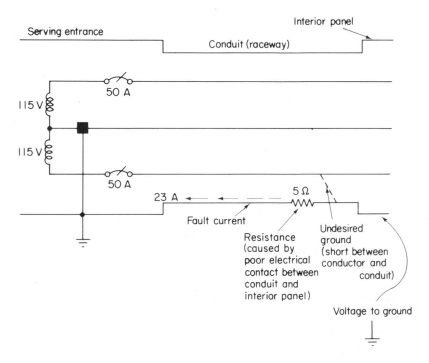

FIGURE 4-14　Effect of poor electrical contact in equipment.

normal operation of the system (since a properly designed electrical system does not depend upon the raceways or equipment to carry normal currents). However, if there is an accidental ground and a fault current flows, the amount of the fault current will be about 23 A (115/5 = 23).

Assume that the overcurrent device (circuit breaker shown in Fig. 4-14) is rated at 50 A. The 23-A fault current will not be sufficient to trip the circuit breaker, and there will be some voltage to ground from the interior panel. There are three problems created by this condition.

First, there is the obvious *shock hazard*, should anyone touch the interior panel. Next, if the raceway is making very poor contact with the panel, there could be some *arcing* at the point where the panel and raceway touch. The arcing could produce a fire. An even greater fire danger exists because of *heating*. Portions of the raceway and interior panel are dissipating 1645 W (115 × 23 = 1645), thus making the raceway and panel equivalent to a 1600-W electrical heater. Of course, the 5-Ω resistance is not necessarily a practical value and is given here for purposes of illustration. However, the practical problem does exist and has caused fires. For this reason some local codes require bonding (by means of a conductor) between raceways and panels or any other place where there is a possibility of poor electrical contact in the equipment.

4-5 PROBLEMS OF DEFECTIVE ARMORED CABLE GROUNDING

In electrical systems that use armored cables as raceways, the cable armor itself provides the path for fault currents. As a general rule, cable armor has less resistance than conduit (designed for the same number of conductors). This is especially true for smaller cables.

The secret of low resistance in armored cable is a ribbon of aluminum that runs the entire length of the cable. The aluminum ribbon is flat and is positioned within the cable between the insulated conductors and the armor. Since the armor and ribbon are parallel, the overall resistance of the cable from end to end is quite low.

Armored cable must be assembled carefully, particularly where the cable enters outlets, boxes, and so on. Generally, the cable is

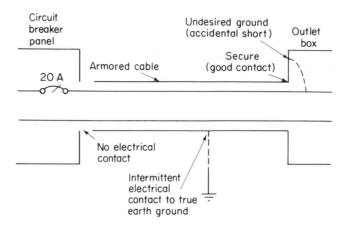

FIGURE 4-15   Effect of poor electrical contact between armored cable and panel, combined with intermittent and accidental contact (undesired grounds).

assembled with clamps to ensure good electrical (as well as mechanical) contact. A problem can develop if the cable is not making good electrical contact. This condition is shown in Fig. 4-15. Here a two-wire system is protected by a 20-A circuit breaker. Armored cable is used between the circuit breaker cabinet and the outlet box. The cable is secure at the outlet but not at the circuit breaker cabinet, and the cable is making intermittent contact with true earth ground (say, a water pipe).

Under the conditions of Fig. 4-15, there is no low-resistance electrical contact between the armored cable or outlet and ground beyond the circuit breaker cabinet. Should there be an accidental ground between one conductor and the outlet box, either one of two conditions will occur. If the intermittent contact from the armored cable to ground is one of low resistance and remains long enough, the circuit breaker will open. If the contact is truly intermittent, there will be arcing at the contact. In any event, the armored cable and outlet box are "hot" and can be a shock hazard.

4-6 DETECTORS FOR UNGROUNDED SYSTEMS

As discussed in Section 4-1, there are no NEC requirements for grounding three-phase three-wire systems. For that reason, ground detector systems are often used with such systems. These ground

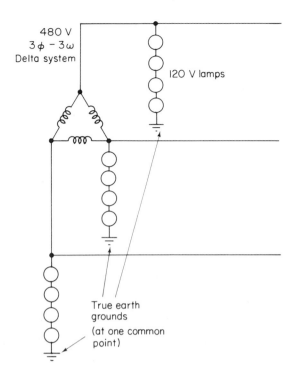

FIGURE 4-16   Simple ground detector for ungrounded 3-phase, 3-wire
distribution system.

detectors indicate if any one of the three conductors is accidently
grounded. There are a number of schemes used for ground detectors.
One of the simplest ground detectors for 440-V three-phase systems
is shown in Fig. 4-16. The detector consists of twelve 115-V lamps
connected in groups-four from each conductor to ground.

    With all three conductors ungrounded, all the lamps will glow
with the same brilliance. (Note that this is not full brilliance, since
each lamp has about 60 to 65 V across it.) If any one conductor becomes grounded, the group of lamps connected to that conductor will
go out. If one conductor has a high resistance to ground, the corresponding lamps will become dim.

    The ground detector of Fig. 4-16 is often used on ungrounded
three-phase systems where there are a number of motors, machines,
controllers, or similar pieces of equipment which have metal surfaces
that are readily accessible. If one conductor of such a system should
contact the metal surfaces, a serious shock hazard exists.

    The detector of Fig. 4-16 can also be used with three-phase four-wire systems (using the appropriate number of lamps). However, such a detector is generally not required, since an overcurrent device will open should one conductor come into contact with a ground. The same is true of single-phase two-wire systems. Of course, as a quick check, it is possible to test two-wire systems with a single lamp. Simply connect the lamp from ground to each of the two conductors, in turn. *Only one conductor should cause the lamp to go out.* If the lamp is on for both conductors, one of the conductors is not properly grounded as required by NEC.

4-7 GROUNDS FOR LIGHTING SYSTEM WIRING

As discussed, the NEC requires that one conductor of all two-wire systems be grounded. When this two-wire system is used for lighting, *the NEC further requires that the grounded conductor be connected to the screw shell of the lampholder,* not to the center terminal.

Figure 4-17 shows both the correct and incorrect grounding arrangement for two-wire lighting systems. In Fig. 4-17a (incorrect), the grounded conductor is connected to the center terminal, with the ungrounded conductor at the screw shell. If a person should accidently touch the screw shell and is simultaneously in contact with true earth ground, he will receive the full 115 V, even though he is not in direct contact with both conductors.

In Fig. 4-17b (correct), the ungrounded conductor is connected to the center terminal, as required by NEC. There is no voltage to ground from the screw shell and thus no shock hazard.

4-7.1 Identifying ground conductors and terminals

The NEC requires that the grounded conductor be identified by a white (or neutral gray) outer covering or insulation. Sometimes the words "ground" or "neutral" will be stamped on the outer cover of the grounded conductor. If the conductor is larger than No. 6 AWG, the identification (white color) need not cover the entire length of the conductor, but it must be at *all terminal ends.* This is usually done with white tape or paint. Any other color, *except green,* can be used for the ungrounded conductors.

The center terminal of lampholders, and the "hot" or ungrounded terminals of receptacles, are identified by brass-colored screws (although the material is not necessarily brass). The screw-shell term-

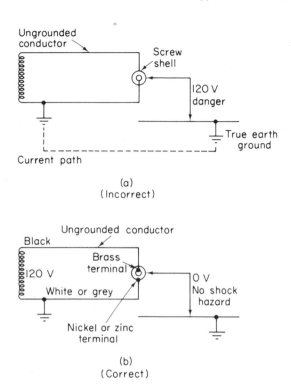

FIGURE 4-17   Correct and incorrect methods of grounding lampholders in lighting system wiring.

inal and grounded receptacle terminal, is silver-, nickel-, or zinc-colored. This is shown in Fig. 4-17. Some lampholders and lighting fixtures have wire leads instead of terminals. In these cases, a white (or neutral gray) wire is used to identify the screw shell.

The proper grounding of typical household lamps brings up an obvious problem. Most house lamps have two-prong plugs that can be reversed in the receptacle without affecting operation of the lamp. There is no way to assure that the screw shell of the lamp is connected to the grounded conductor. This is the advantage of a three-prong plug, as discussed in Section 4-8.

4-8 GROUNDS FOR APPLIANCES AND PORTABLE TOOLS

Just as it is possible for a person to touch the screw shell of a lamp-holder, it is almost impossible to operate household appliances (washing machines, dryer, refrigerator, etc.) or power tools without touching the frames or metal surfaces. Of course, such equipment is designed so that neither the ungrounded or grounded conductors make contact with the metal surfaces. However, a defect (such as worn insulation) could occur where one conductor does touch the metal surfaces. This creates an obvious shock hazard, as shown in Fig. 4-18a.

In recent years this problem has been minimized by NEC requirements that most appliances and portable tools be provided with three-wire cords and three-prong *polarized* plugs, as shown in Fig. 4-18b. The third wire, which must have a green outer cover, provides an equipment ground by connecting the appliance or tool metal surface back to the raceway (through the receptacle box). In turn, the raceway is connected to true earth ground, along with the grounded conductors, at the service entrance. Of course, this does not prevent an accidental ground, but should there be one, the overcurrent device will open. At least there will be no shock hazard, since the metal surface will be at true earth ground.

As shown in Fig. 4-18b, the three-prong plugs are polarized by means of a U-shaped terminal for the equipment ground, or green, conductor. These plugs do not fit the old style two-terminal receptacles. This mismatch problem can be overcome, however. In some appliances and portable tools, the green equipment ground conductor is provided with a lug-type fastener, and is separate from the other two current-carrying conductors (which are fitted with a two-prong plug). The terminal lug on the green conductor is attached to the raceway by means of the screw that holds the receptacle cover plate. This is shown in Fig. 4-18c.

Likewise, there are adapters to solve the mismatch problem. These adapters have a three-prong receptacle, a two-prong plug, and a ground wire with a terminal lug. The adapter mounts into the wall receptacle by means of the two-prong plug. The adapter's ground wire is connected to ground by means of the cover-plate screw. The appliance or tool cord (with a new-style three-prong plug) is then connected to the adapter's three-prong receptacle. This is shown in Fig. 4-18d.

Virtually all new-house wiring uses three-prong polarized receptacles, so the problem should be minimized in the future. When large appliances (such as dryers and ranges) are operated from 230 V with three conductors, the neutral conductor can be connected to

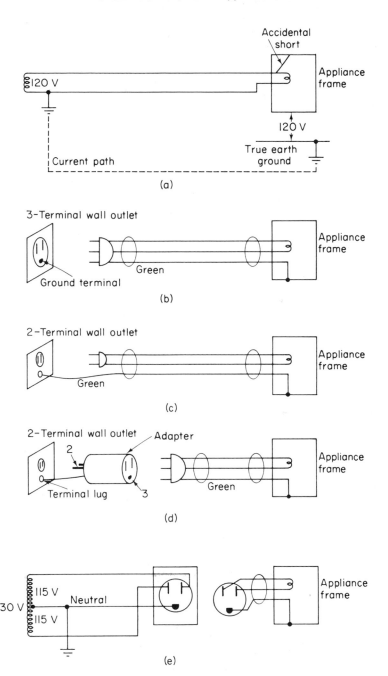

FIGURE 4-18   Various ground systems for appliances and portable tools.

metal surfaces or frames of such appliances, as shown in Fig. 4-18e. This arrangement is not in common use, however.

4-9 SWITCHES AND OVERCURRENT PROTECTION IN GROUNDS

The NEC generally requires a "solid neutral" electrical system. This means that the grounded conductor of any system must not be provided with an overcurrent device. Further, *the NEC prohibits using a switch in the grounded conductor*-except where *all the conductors can be* opened by a single switch (two-pole switch for two-wire systems, three-pole switch for three-wire systems, etc.).

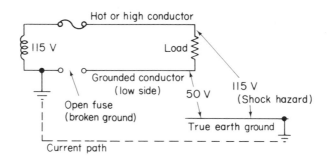

FIGURE 4-19   Effect of double fusing a 2-wire distribution system (NEC violation).

Figure 4-19 shows what could happen if both conductors of a two-wire system are provided with fuses (a condition known as *double fusing*). If the fuse in the ground conductor is opened by an overcurrent, the high or "hot" side of the load is at 115 V with respect to ground, and a shock hazard can exist. Going further, the ground or low side of the load can also be at some voltage with respect to ground if the load does not drop the entire 115 V. For example, if a person should touch the low side of the load and the resultant current is sufficient to produce a drop of 65 V across the load, the person will receive 50 V.

Figure 4-20 shows incorrect switching and two alternative methods of correct switching for a two-wire system. Figure 4-20a shows what can happen if a two-wire system is provided with a single-pole switch in the grounded conductor (incorrect). When the switch is opened, power is removed to the load. However, the high side of the load is

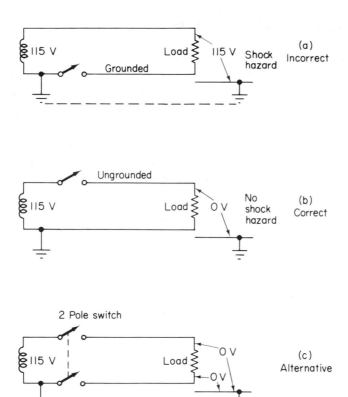

FIGURE 4-20   Correct and incorrect methods of switching in 2-wire distribution system.

still at 115 V in respect to ground, and a shock hazard can exist (as described for double fusing).

In Fig. 4-20b a single-pole switch is correctly installed in the ungrounded conductor of a two-wire system. Power is removed when the switch is opened. Both the high and low sides of the load are at zero volts with respect to ground.

In Fig. 4-20c a two-pole switch is installed in the conductors of a two-wire system. Power is removed when the switch is opened and *both sides of the load* are at zero volts. Of course, a two-pole switch is not required for a two-wire system, but this does not violate NEC requirements.

As long as all ungrounded conductors are opened simultaneously by a switch or a circuit breaker, there will be no shock hazard (and no NEC violation).

4-10 GROUND FAULT CIRCUIT INTERRUPTERS

A *ground fault circuit interrupter* (GFCI) is "a device whose function is to interrupt the electric circuit to the load when a fault current to ground exceeds a predetermined value that is less than that required to operate the overcurrent protective device of the supply circuit."

The Underwriters' Laboratories have adopted 5 milliamperes (mA) as the trip level at which devices designated as "Class A" must operate. This classification is for GFCI units intended for general use as personnel protectors. The latest NEC incorporates several requirements for GFCIs. These requirements include GFCI units for construction sites, outdoor outlets in residential occupancies, boat harbor circuits, receptacles near swimming pools, storable swimming pool circuits, and underwater lighting fixtures in permanently installed pools.

Several different types of equipment have been developed to provide this. Four types of GFCI are in common use. These types include permanently installed receptacles (usually weatherproof for outdoor use), portable receptacles (for indoor use), fixed mounting units, and circuit breakers (which include ground fault protection in addition to normal short-circuit and overload protection).

4-10.1 GFCI circuit breakers

Figure 4-21 illustrates a typical circuit breaker that includes ground fault protection. Installation of such circuit breakers varies slightly from that of standard circuit breakers. Both the "hot" and neutral conductors of the circuit to be protected are connected to the circuit breaker, providing the means to pass both the "hot" and neutral current through a differential current transformer sensor located within the circuit breaker. Most 120-V GFCI units presently manufactured in the United States are designed to operate on the *current imbalance* in the "hot" and neutral conductor that results when a ground fault occurs. An imbalance of 5 mA is established as the tripping level for "people protector" GFCI units.

In a typical GFCI circuit breaker, the line-side "hot" connection is made by plugging or bolting the breaker to a panelboard bus structure. A wire "pigtail," which is the line neutral connection, is connected to the panelboard neutral assembly. Typical circuit breaker wiring with and without GFCI is shown in Figs. 4-22 and 4-23.

FIGURE 4-21   Typical GFCI unit
(Courtesy Square D Company).

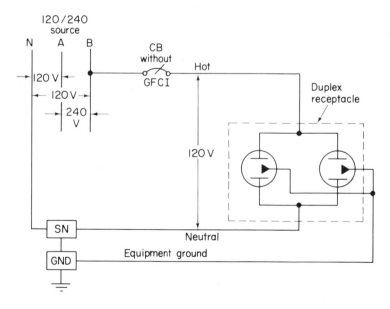

FIGURE 4-22   Typical circuit breaker wiring without GFCI.

As shown in Fig. 4-22, circuit breakers without GFCI protection are usually connected to the "hot" wire only. As shown in Fig. 4-23, which is the wiring of a Square D Qwik-gard circuit breaker on a 120-V

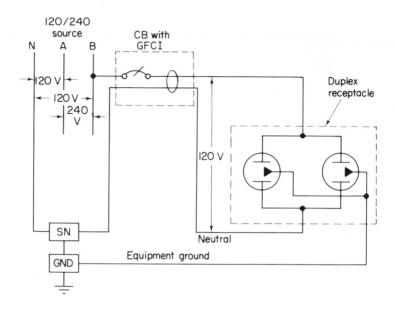

FIGURE 4-23   Typical circuit breaker wiring with GFCI.

duplex receptacle circuit, the breaker is connected to both the "hot" and neutral lines. Figure 4-24 shows the circuitry of the Qwik-gard circuit breaker. All components and connections shown are within the breaker.

In Fig. 4-25 a ground fault is shown in a typical branch circuit. A path is established for current to flow back to the source by means

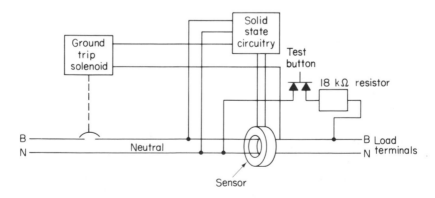

FIGURE 4-24   Basic internal circuitry of Square D Company QWIK-GARD™ circuit breaker with GFCI.

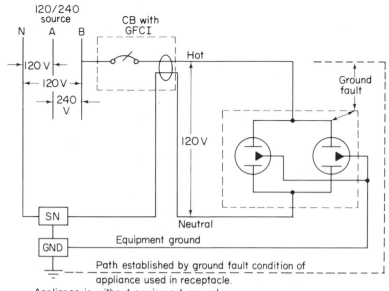

FIGURE 4-25   Effect of ground fault on the sensor within the GFCI circuit breaker.

other than through the branch neutral conductor. Since a circuit breaker that includes GFCI is protecting the circuit, a sensor method for monitoring the current balance is provided. When the sensor detects an imbalance of 5 mA or more, the GFCI solid-state circuitry energizes the breaker ground trip solenoid, and the breaker opens the circuit.

It should be noted that GFCI circuit breakers are independent of the equipment ground conductor (Section 4-2) and can be used on "older" installations, which do not have the equipment ground provisions, as well as on up-to-date installations, which do include the ground conductor (see Section 4-8).

Since GFCI units operate on a current imbalance between the normal current-carrying conductors of the protected circuit, GFCI units *provide no protection to the user against contacts that are the same as normal load conditions.* For example, if a person insulated from (or in poor contact with) ground should touch both the "hot" and neutral conductors of a 120-V circuit, this will appear as a normal load condition to the GFCI sensor, and the GFCI will not trip. The

GFCI will trip only if the undesired contact is made between the "hot" conductor and ground.

4-10.2 Testing GFCI units

The test circuit shown in Fig. 4-24, required by Underwriters' Laboratories, produces a 6.7-mA ground fault within the breaker when the test button is pressed. (The circuit breaker must be operated at the rated 120 V to produce the full 6.7 mA.) The level of ground fault (amount of fault current) is set by the test circuit resistor. When the test button is pressed, and the simulated ground fault current flows, the breaker will trip just as in an actual ground fault condition. Thus operation of the entire circuit breaker is checked. After test, the circuit breaker must be reset in the normal manner.

The UL standards for listing GFCI devices include tests on operation of the GFCI under many conditions. Specific performance requirements must be met concerning the operation as affected by various conditions, such as voltage changes, ambient temperature, humidity, flooded conduit, and accidental grounding of the neutral conductor of a protected circuit. For example, the typical GFCI circuit breaker must operate over a temperature range from -35°C to +66°C, plus or minus 2°, and from 120 to 132 V, for a device rated at 120 V.

4-10.3 Examples of ground faults that will trip GFCI units

Figures 4-26 and 4-27 illustrate occurrences that will *appear as ground fault* conditions and will result in the GFCI device opening the circuit.

In the circuit of Fig. 4-26, current will flow from the source, through the circuit breaker and sensor, to the receptacle. From there current will flow through the faulty appliance and back to the source through the equipment ground conductor.

The current path in the circuit of Fig. 4-27 is from the source, through the circuit breaker and sensor, to the receptacle. From there the current flows through the normal load connected to the receptacle and back to the source through both the neutral and the unintended path established by the neutral being grounded on the load side of the circuit breaker. In such a case, current will split at random between the normal neutral and unintentional path through ground.

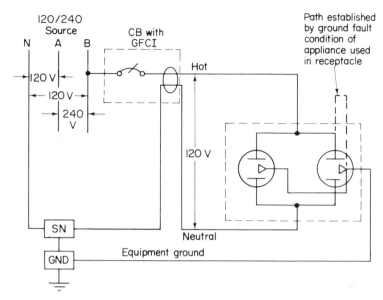

Appliance is faulty and has equipment ground

FIGURE 4-26   Effect of faulty appliance (with equipment ground) on GFCI circuit breaker.

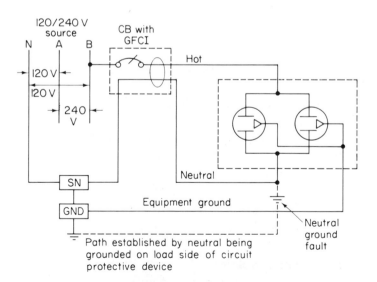

FIGURE 4-27   Effect of ground fault that occurs on neutral conductor at the load side of the circuit breaker.

4-10.4 Problem areas with GFCI devices

As with any new development, equipment is continually improved or changed based on increased use and field experience. Installation by persons unfamiliar with operating requirements of the GFCI, or not following NEC wiring rules, can present some problems.

Proper installation of GFCI units is most important. Figure 4-28 illustrates an *improper installation* that could result from carelessness or from wiring done by an unqualified person. Such a condition can result when adding a subpanel (for branch circuits, not at the service entrance) to an existing system. In such panels the neutral is not grounded, as discussed in Section 4-3.

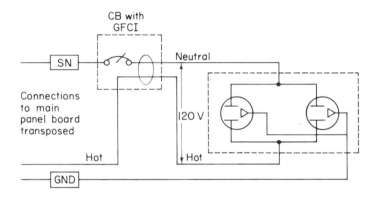

FIGURE 4-28   Improper installation of GFCI circuit breaker on sub-panel where connections to main panelboard are transposed.

In the improper installation of Fig. 4-28, the subpanel appears only as a 120-V load, the same as other normal loads (lamps, appliances, etc.) that can operate with *either* connection made to the "hot" and the corresponding connection made to the neutral. However, in the circuit of Fig. 4-28, the GFCI will open under a ground fault condition *but will not protect the user.* The breaker will open the neutral conductor, leaving the person exposed to the "hot" conductor. *GFCI devices must be installed so as to interrupt all the "hot" conductors of the protected circuit.*

Common-neutral problems. Figure 4-29 shows the use of a *common neutral* for three-wire branch circuits, using breakers without GFCI. The same conditions exist whether the loads are separate duplex receptacles or single duplex receptacles that are "split-wired." (Split

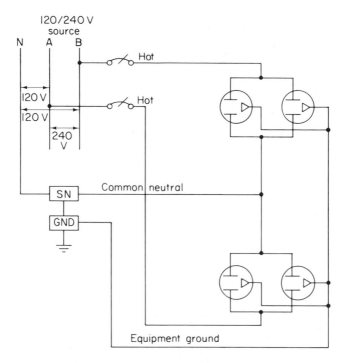

FIGURE 4-29   Typical *common-neutral* wiring of 3-wire system; cannot be used with GFCI circuit breakers.

wiring is accomplished in duplex receptacles when two single re-ceptacles are obtained by removing an interconnecting link between the two "hot" connections on the duplex arrangement.) The common neutral wiring technique is sometimes used for economic reasons, since some conductors can be eliminated from the branch circuit wiring.

Since single-pole GFCI devices function on the current imbalance between the two conductors passing through the sensor, *the com-mon neutral wiring technique cannot be used.*

Figure 4-30 shows the use of two single-pole GFCI circuit breakers in a wiring system similar to the common neutral arrangement. In the circuit of Fig. 4-30, the load neutral connection *has not* been made to breaker 2.

When a load is applied only at receptacle 1, both the "hot" and neutral currents are carried through breaker 1 and no problems exist. However, as soon as a load is applied at receptacle 2, breakers 1 and 2

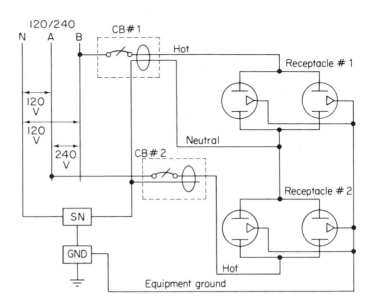

FIGURE 4-30   Improper wiring of two GFCI circuit breakers with common neutral.

will both trip, since both sensors will see unbalanced currents. Breaker 2 sensor sees only the "hot" current to receptacle 2. Breaker 1 sensor sees the "hot" current to receptacle 1 and the neutral current from both receptacles 1 and 2.

Tying the two breaker load neutral terminals together in the panel will not solve the problem, because a load at either receptacle 1 or 2 will unbalance both sensors, since the neutral current will split at random through the two sensors. In some cases neutral conductors for more than one branch circuit are combined under one terminal or connector in a junction box. *This technique cannot be used where GFCI is involved*, since such a connection results in parallel return paths for each of the branch circuits involved (causing an imbalance in the GFCI sensor).

A general rule to remember in GFCI devices is that all conductors (except equipment ground) for a circuit must pass through a single sensor, and these conductors cannot be shared by any other circuit. If the currents passing through a GFCI sensor do not vectorially total less than the operating current of the GFCI, the device will open the circuit.

4-10.5 Accidental or false tripping

All GFCI devices can be tripped accidently. The following is a summary of the major causes for such false tripping.

GFCI devices may trip as a result of current leakage to ground. Such leakage can be caused by defective appliances or by the *accumulative current to many good appliances* operating on the same circuit (where the leakage is within permissible limits). Leakage can also result from situations such as receptacles becoming wet. In certain cases of current leakage, the circumstances would be hazardous, and the GFCI will open the circuit for protection of the user. However, since a GFCI has no way of recognizing circumstances, tripping many occasionally occur when the conditions are not actually hazardous. Changes are being made in equipment designs to reduce leakage current so that the number of accidental trips can be reduced.

When considering possible accumulative current leakages and other causes of GFCI opening of circuits, it is obvious that GFCI protection should be located in individual branch circuits and not in the main or feeder circuits. With a GFCI in a branch circuit, a problem in one circuit will not shut down the entire area or all the power in a residence. When only the branch circuit is opened, the other circuits remain in operation (and can provide lighting to find the trouble). Likewise, since GFCI circuit breakers should be tested (by means of the push-to-test provision), the use of GFCI devices in branch circuits results in a minimum of disturbance, since the entire panel is not interrupted.

GFCI devices are sometimes installed on circuits other than those specifically required by the NEC. False trippings have been reported on circuits where high-voltage spikes occur during the opening of inductive circuits with relays, contactors, etc., (such as the motor control circuits described in Chapter 8). Most problems of this type can be solved by the addition of a capacitor to the inductive circuit being opened. The capacitor must be of proper value to limit the voltage spike to a level the GFCI can withstand. Such problems must be handled on an individual basis. With information such as the magnitude and duration of the voltage spike for a particular device, most manufacturers of GFCI units can make a recommendation for the proper capacitor to be used.

Electrical storms can also produce voltage spikes on the line that will trip a GFCI. However, this is a rare occurrence and can generally be neglected.

Chapter 5

Applications of Transformers

For many years transformers have been used extensively in the design of industrial or heavy-duty electrical wiring systems. The main purpose of such transformers is to provide several different voltages, for use by different loads, from a single voltage source. A classic example is where an industrial plant requires 480-V three-phase power for electrical motors, and 120-V single-phase for lighting. Were it not for transformers, it would be necessary to supply two separate electrical services to the plant. With transformers it is possible to supply only the 480-V three-phase power to the service entrance and distribute this power to the motors. A transformer at or near the service entrance (or near the load) is connected to convert the 480-V three-phase power into 120-V single-phase and distribute the single-phase power to the lighting.

Until recently, transformers were not used extensively in residential electrical wiring. About the only exception was the use of transformers to convert 120 V into some low voltage (usually 6 or 12 V) for doorbells. Today there is an increasing use of low-voltage electrical systems for residential applications. All these systems require transformers. In this chapter we shall discuss both the conventional transformer systems and the newer low-voltage wiring systems.

5-1 TRANSFORMER BASICS

5-1.1 Voltage, current, impedance, and turns ratios

The voltage, current, and impedance ratios of transformers are dependent upon the *turns ratio* of the primary and secondary windings. The calculations for voltage, current, and impedance ratios of transformers are shown in Fig. 5-1.

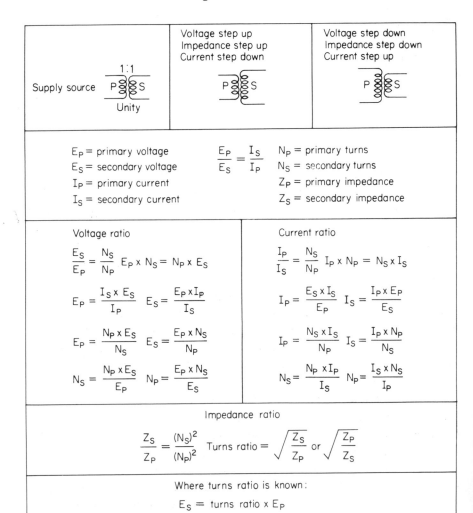

FIGURE 5-1   Calculations for voltage, current, and impedance ratios of transformers.

The *voltage* across a transformer is stepped up when the secondary has more turns than the primary. The voltage is stepped down when the primary has more turns than the secondary.

The *current* across a transformer is stepped up when the primary has more turns than the secondary. The current is stepped down when the secondary has more turns than the primary.

When the primary and secondary have the same number of turns, the transformer is said to have *unity* or 1:1 *coupling*.

The *impedance* across a transformer is stepped up when the secondary has more turns than the primary. The impedance varies as the *square* of the turns ratio.

The following are some examples of how these relationships can be used to calculate voltage and currents in electrical systems with transformers.

Assume that a transformer must be selected to supply 30 A and 120 V at its secondary winding. The available source is 480 V. Find the primary current using the equations of Fig. 5-1.

The primary current, I_P, is found by

$$I_P = \frac{E_S \times I_S}{E_P} \quad \text{or} \quad \frac{120 \times 30}{480} = 7.5 \text{ A}$$

Assume that another transformer is connected to the same source. The primary has 500 turns, the secondary has 200 turns. What voltage is available at the secondary?

Secondary voltage, E_S, is found by

$$E_S = \frac{E_P \times N_S}{N_P} \quad \text{or} \quad \frac{480 \times 200}{500} = 192 \text{ V}$$

Assume that the primary of the same 480/192-V transformer is connected to an unknown source. The secondary is connected to an 8-A load. Find the primary current.

$$I_P = \frac{N_S \times I_S}{N_P} \quad \text{or} \quad \frac{200 \times 8}{500} = 3.2 \text{ A}$$

Assume that a transformer shows 240 V on the primary and 80 V at the secondary. What is the impedance ratio (primary to secondary)?

$$\frac{E_S}{E_P} = \frac{N_S}{N_P}$$

Therefore,

$$\frac{E_P}{E_S} = \frac{N_P}{N_S} = \frac{240}{80} = \frac{3}{1}$$

$$\frac{Z_P}{Z_S} = \frac{N_P^2}{N_S^2} = \frac{3^2}{1^2} = 9$$

5-1.2 Transformer winding polarity markings

In theoretical problems such as shown in Fig. 5-1, transformer windings are marked as primary and secondary. Generally, the primary is connected to the source, with the secondary connected to the load. Usually, there is a step up or a step down in turns between primary and secondary. An isolation transformer is an exception to this. Isolation transformers have the same primary and secondary windings and are used to eliminate any direct contact between primary and secondary circuits (source and load).

In practical applications, transformer windings are generally marked as to high and low voltage rather than primary and secondary. *Low-voltage* windings are typically marked with an X. X_1 and X_2 are used to identify the opposite terminals of one winding. If there are more low-voltage windings on the same transformer, the second winding is marked X_3 and X_4, the third winding is marked X_5 and X_6, and so on. *High-voltage* windings are marked with an H (H_1 and H_2, H_3 and H_4, etc.).

The numbers used to identify transformer windings are also significant. All even-number terminals will have the same instantaneous polarity. That is, assuming an alternating current, all even-number terminals will swing positive at the same time. All odd-number terminals will swing negative at the same instant. All even-number terminals are in phase with each other and out of phase with all odd-number terminals.

On some small transformers the terminals are not numbered; rather, the polarity or phase is marked by a dot. All terminals with a dot have the same phase and are out of phase with all terminals that do not have a dot.

Whatever system is used, the polarity markings *must be observed* when connecting transformer windings. For example, assume that a transformer has two identical secondary windings marked H_1 and H_2,

H_3 and H_4, and each winding produces 120 V. If the odd terminals are connected together (H_1-H_3, H_2-H_4), the output will be 120 V. However, if an odd terminal is connected to an even terminal (say, H_1-H_4) and the output is taken from the opposite terminals, the output will be 240 V (or the sum of the two 120-V windings). The same results are produced when H_2 is connected to H_3 and the output is taken from H_1 and H_4.

No matter how transformer windings are connected, *never connect any winding to a voltage higher than the rating for that particular winding.* An exception to this is where two (or more) windings are connected in series. For example, using the same transformer with 120-V windings, if H_1 is connected to H_4, H_2 and H_3 could be connected to a 240-V source (provided that the current ratings were not exceeded).

5-1.3 Transformer power ratings

In addition to voltage and frequency, transformer windings are rated as to power. Generally, transformers are rated in voltamperes (VA) or kilovoltamperes (kVA) rather than in watts. Since the power factor of the load may be unknown, a rating in watts would be difficult to use.

The VA or kVA rating of a transformer shows the safe limit of load that can be connected (generally to the secondary circuit). For example, if a transformer is rated at 3000 VA (or 3 kVA) and the load is connected to a 120-V winding, the maximum safe current in that winding is $3000/120 = 25$ A.

The VA or kVA rating also applies to the source (or primary) winding. For example, assume that the same 3-kVA transformer has a 240-V source winding. Then the maximum safe current in that winding is $3000/240 = 12.5$ A.

When a transformer has more than one coil in either the primary or secondary, the power rating of the transformer must be divided by the number of windings in each section. (In practical applications the windings in each section of most multiwinding transformers have the same voltage rating. This permits the windings to be operated in parallel, as discussed in later sections of this chapter.)

Assume that the same 3-kVA transformer had two 240-V primary (or source) windings, and two 120-V secondary (or load) windings.

Then the maximum safe current for the 240-V windings is 3000/2 = 1500; 1500/240 = 6.25 A. The maximum safe current for the 120-V windings is 1500/120 V = 12.5 A.

5-1.4 Transformer losses and efficiency

All transformers have some losses. That is, all transformers introduce some power loss into the electrical system. Stated another way, the output power from any transformer is always somewhat less than the input power. The lower the output power, in relation to input power, the lower the efficiency (as indicated by a lower percentage of efficiency).

There are two common types of transformer loss: *copper loss* and *core loss*. *Copper loss*, also known as I^2R loss, resistance loss, or (possibly) heat loss, is the result of resistance in the transformer windings. All windings have some resistance that requires power to overcome, which results in a loss of power. Copper loss varies as the square of the load current. For example, assume a load current of 3 A and a winding resistance of 5 Ω. This produces a copper loss of I^2R, or $3^2 \times 5 = 45$ W (or 45 VA, assuming a unity power factor). Now assume that the load current is increased to 7 A. This produces a copper loss of $7^2 \times 5 = 245$ W.

Core loss occurs because a small amount of power is necessary to force the flux in the transformer's magnetic core material to alternate at the frequency of the source voltage. Core loss is always present and is independent of transformer loading. However, core loss is calculated on the basis of full-rated voltage. For example, if a transformer is rated at 240 V (primary or source) and is connected to a 120-V source, the core loss will be less.

Copper loss and core loss must be combined to find the total loss of the transformer. In turn, the combined copper and core losses are compared with the transformer's power rating to find efficiency. For example, assume that a 30-kVA transformer has a core loss of 300 W and a copper loss of 400 W. The percentage of efficiency is found by

300 W (core) + 400 W (copper) = 700 W combined loss

input power = 30,000 W + 700 W = 30,700 W

output power = 30,000 W

efficiency = (output/input) $\times$ 100 = (30,000/30,700)
$\times$ 100 = 97.7%

Typically, copper losses and core losses are both less than 2 per cent of the transformer's power rating. Thus the combined losses are less than 4 per cent of the power rating. The efficiency or the combined losses or both are sometimes given on the transformer's nameplate. The procedures for measurement of transformer losses are given in Section 5-1.6.

5-1.5 Transformer regulation and per cent of impedance

All transformers have some voltage-regulating effect. That is, the output voltage of a transformer tends to remain constant with changes in load. Regulation is usually expressed as a percentage and is equal to

$$\% \text{ of regulation} = \frac{\text{no-load voltage} - \text{load voltage}}{\text{no-load voltage}} \times 100$$

Some transformers show good regulation (a low percentage). Other transformers provide very poor regulation (a high percentage). For example, assume that the no-load voltage at a transformer output (secondary) is 120 V, and the load voltage (at full-rated load) is 115 V. The percentage of regulation is (120 − 115)/120 X 100 = 4.16 per cent. If the load voltage dropped to 110 V (poorer regulation), the percentage of regulation is (120 − 110)/120 X 100 = 8.33 per cent.

Transformer regulation is also expressed, and measured, as a percentage of impedance. This measure (sometimes shown as "per cent Z" or "%Z" on transformer nameplates) involves the primary (or source) voltage necessary to force rated current to flow through a short-circuited secondary (or load) winding, compared to the rated primary voltage, or

$$\% \text{ of impedance} = \frac{\text{primary voltage (with secondary shorted)}}{\text{rated primary voltage}} \times 100$$

For example, assume that a transformer with a primary rated at 240 V requires 12 V to produce the rated current in a short-circuited secondary winding. The percentage of impedance is (12/240) X 100 = 5 per cent. If the voltage must be increased to 24 V, the percentage of impedance is (24/240) X 100 = 10 per cent.

5-1.6 Transformer testing

Although it is not necessarily the job of the electrical system designer to test transformers, the need for such tests often arises in practical work. The tests most often needed include checking phase relationships, checking polarity markings, measuring copper loss, measuring core loss, measuring transformer efficiency, measuring percentage of impedance, measuring transformer regulation, measuring impedance ratio, and winding balance. The following sections describe the procedures for such tests.

5-1.6.1 Checking phase relationships. When two supposedly identical transformers must be operated in parallel (at the primary or source), and the transformers are not marked as to phase or polarity, the phase relationship of the transformers can be checked using a voltmeter and a power source. The test circuit is shown in Fig. 5-2.

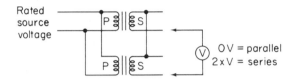

FIGURE 5-2   Checking transformer phase relationships.

The transformers are connected in proper phase relation for parallel operation if the meter reading is zero or is very low. The transformers are connected for series operation (secondaries in series, adding) if the output voltage is double that of the normal secondary output for one transformer.

If the meter reading is zero or very low and series operation is desired, reverse either the primary or the secondary terminals (not both) of one transformer (not both transformers).

If the meter reading is double the rated secondary voltage and parallel operation is desired, reverse either the primary or secondary terminals (not both) of one transformer (not both transformers).

If the meter indicates some secondary voltage, far below that of the rated voltage (but not zero), it is possible that the transformers are not identical. One transformer has a greater output than the other. This condition will result in considerable *local current* flowing in the secondary winding and will produce a power loss (if not actual damage to the transformers).

Thus it will be seen that the test circuit of Fig. 5-2 not only checks phase relationships but also indicates matching of the transformers.

5-1.6.2 Checking polarity markings. Most transformer windings are marked as to polarity (or phase), as discussed in Section 5-1.2. However, the markings may not be clear, or some new system of identification may be used. When this occurs it is possible to identify the polarity markings using only a voltage source and a voltmeter.

When a transformer has only one primary and one secondary (or one high and one low voltage) winding, the problem of identifying the polarity markings is relatively simple. From a practical standpoint there are only two problems of concern with single-winding transformers: the relationship of the primary to the secondary, and the relationship of the markings on one transformer to those on another.

The phase relationship of primary to secondary can be determined using the test circuit of Fig. 5-3. First connect one primary terminal to one secondary terminal. Apply a voltage to the primary (or low-voltage section). It need not be the full rated voltage. In fact, a lower voltage is often convenient. It is easy to separate the high-voltage section from the low-voltage section. The low-voltage section has larger wires, since it must carry a greater current.

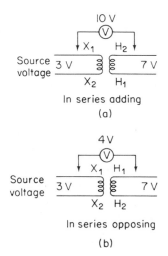

In series adding
(a)

In series opposing
(b)

FIGURE 5-3   Checking transformer polarity markings.

With a test voltage applied to the low-voltage section, measure the voltage across X_1 and H_2 and then across X_1 and H_1. Assume that there is 3 V across the low-voltage section and 7 V across the high-voltage section.

If the windings are as shown in Fig. 5-3a, the 3 V will be added to the 7 V and will appear as 10 V across X_1 and H_2. If the windings are as shown in Fig. 5-3b, the voltages will oppose and will appear as

4 V (7 -3) across X_1 and H_1. In either case the phase relationship between primary and secondary is established.

The phase relationship of markings on one transformer to those on another can be found using the test circuit of Fig. 5-4. Assume that there is a 3-V output from the high-voltage winding of transformer A and a 7-V output from transformer B. If the markings are consistent on both transformers, the two voltages will oppose and 4 V will be indicated. If the markings *are not* consistent the two voltages will add, resulting in a 10-V reading.

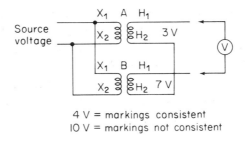

4 V = markings consistent
I0 V = markings not consistent

FIGURE 5-4   Checking polarity markings of one transformer against those of another transformer.

When a transformer has multiple windings, the problem of identifying the polarity markings is more complex. Not only is it necessary to establish the phase relationship between primary and secondary windings, but it is also necessary to find phase relationship between the primary windings (as well as between the secondary windings).

The phase relationship of multiwinding transformers can be found using the test circuit of Fig. 5-5, which is for a transformer with two primary (240 V) and two secondary (120 V) windings. The same procedure can be used for three-phase transformers, except that one additional set of windings (primary and secondary) must be checked. The following steps describe the test procedure.

1. Connect any winding to the proper voltage according to the nameplate information, as shown in Fig. 5-5.
2. Arbitrarily assign H_1 and H_2 to this source winding (or X_1 and X_2 if the winding is of low voltage).
3. Check the other three voltages. They should be as shown in Fig. 5-5 (one 240-V and two 120-V windings).

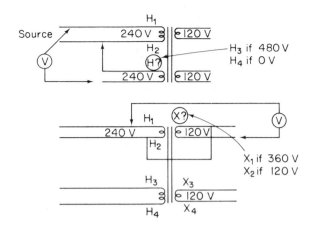

FIGURE 5-5   Checking polarity markings on multi-winding transformers.

4. Connect one lead of the other 240-V winding to H_2 of the source winding.
5. If the voltmeter indicates 480 V, the voltages are additive, and the top terminal is odd-numbered or H_3.
6. If the voltmeter indicates 0 V, the voltages are opposing, and the terminal is even-numbered or H_4.
7. If the transformer is three-phase, repeat the procedure for the third high-voltage winding (H_5 and H_6).
8. Repeat the procedure with each of the 120-V windings. If the voltmeter reads the sum of 240 and 120 V, or 360 V, the common terminal of the 120-V winding is X_1. If the voltmeter shows $240 - 120$ V, or 120 V, the terminal is X_2. Terminals X_3 and X_4 (as well as X_5 and X_6) can be found in the same way.

5-1.6.3 Measuring copper loss. Copper loss can be measured using the test circuit of Fig. 5-6. A wattmeter is used in the circuit of Fig. 5-6a to measure input power. If a wattmeter is not available, the circuit of Fig. 5-6b can be used. However, the circuit of Fig. 5-6b will indicate the copper loss in VA (or kVA) rather than watts.

1. With either circuit, increase the variable voltage source from zero until the ammeter in the low-voltage (short circuit) winding indicates the rated current for that winding. Typically, the variable voltage source will be very low in relation to the rated winding voltage.

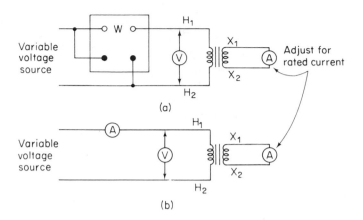

FIGURE 5-6   Measuring transformer copper loss.

2. For the circuit of Fig. 5-6a, simply note the wattmeter reading to determine the copper loss.
3. For the circuit of Fig. 5-6b, multiply the variable voltage indication by the primary or input ammeter reading to find the copper loss.

5-1.6.4 Measuring core loss. Core loss can be measured using the test circuit of Fig. 5-7. A wattmeter is used in the circuit of Fig. 5-7a to measure input power. The circuit of Fig. 5-7b indicates the core loss in VA (or kVA).

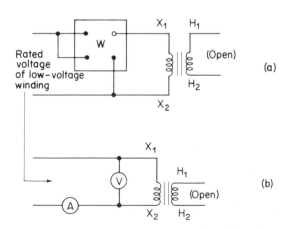

FIGURE 5-7   Measuring transformer core loss.

1. With either circuit, apply the rated voltage to the low-voltage winding. All other windings must be open circuit.
2. For the circuit of Fig. 5-7a, simply note the wattmeter reading to find the core loss.
3. For the circuit of Fig. 5-7b, multiply the voltmeter reading by the ammeter reading to find the core loss.

5-1.6.5 Measuring transformer efficiency. Transformer efficiency is determined by adding the core loss and copper loss to the rated output to find the required input, as discussed in Section 5-1.4.

5-1.6.6 Measuring percentage of impedance. The percentage of impedance can be measured using the test circuit of Fig. 5-8.

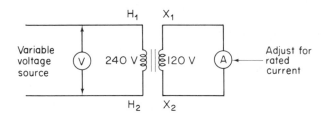

FIGURE 5-8   Measuring transformer efficiency.

1. Increase the variable voltage source from zero until the ammeter in the low-voltage (short-circuit) winding indicates the rated current for that winding. Typically the variable voltage source will be very low in relation to the rated winding voltage.
2. The percentage of impedance is found when the variable source voltage is divided by the rated voltage for the high-voltage winding and the result multiplied by 100. For example, assume that 6 V must be applied to a 240-V high-voltage winding to produce an ammeter reading (in the low-voltage winding) equal to the rated winding current. Then the percentage of impedance is $(6/240) \times 100 = 2.5$ per cent.

5-1.6.7 Measuring transformer regulation. Transformer regulation can be measured using the test circuit of Fig. 5-9. The required load, shown as resistance R_1, must be of a value that will produce full-rated current flow in the secondary or output winding. For example, if the winding is rated at 120 V and 6 A, the value of R_1 is $120/6 = 20\ \Omega$.

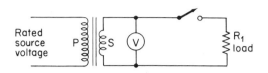

FIGURE 5-9   Measuring transformer regulation.

1. Apply the rated voltage to the input winding.
2. Measure the secondary output voltage without a load.
3. Apply the load (close the switch) and measure the secondary output voltage with the load.
4. Find the percentage of regulation using the information of Section 5-1.5.

5-1.6.8 Measuring impedance ratio. The impedance ratio of a transformer is the square of the winding ratio. Impedance ratio should not be confused with percentage of impedance.

If the winding ratio of a transformer is 15:1, the impedance ratio is 225:1 (15 X 15). Any impedance value placed across one winding will be reflected onto the other winding by a value equal to the impedance ratio. For example, assume an impedance ratio of 225:1 and an 1800-Ω impedance placed on the primary. The secondary will then have an impedance of 8 Ω. Likewise, if a 10-Ω impedance is placed on the secondary, the primary will have a reflected impedance of 2250 Ω.

Impedance ratio is related directly to turns ratio (primary to secondary). However, turns ratio information is not always available, so the ratio must be calculated using a test circuit as shown in Fig. 5-10.

Measure both the primary and secondary voltages. The rated volt-

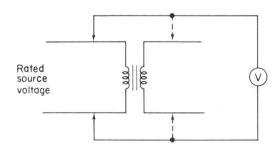

FIGURE 5-10   Measuring transformer impedance ratio.

ages should be used for the test. However, lower-than-rated voltages are acceptable for test purposes.

The *turns ratio* is equal to one voltage divided by the other. The *impedance ratio* is the square of the turns ratio. For example, assume that the primary shows 120 V, with 24 V at the secondary. This indicates a 5:1 turns ratio and a 25:1 impedance ratio.

5-1.6.9 Measuring winding balance. Center-tapped transformers are sometimes used in 240/120-V three-wire single-phase systems. In other cases a transformer with two 120-V windings is used. In either case there is always some imbalance between the windings or on both sides of the center tap. This is, the turns ratio and impedance are not exactly the same on both sides of the center tap (or for both windings). The imbalance is usually of no great concern unless the imbalance is severe.

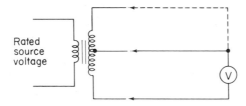

FIGURE 5-11   Measuring transformer winding balance.

It is possible to find a large imbalance by measuring the dc resistance on either side of the center tap. However, it is usually more practical to measure the voltage on both sides, as shown in Fig. 5-11. If the voltages are equal, the transformer winding is balanced. If a large imbalance is indicated by a large voltage difference, the winding should then be checked with an ohmmeter for shorted turns or a similar failure.

5-2 BASIC TRANSFORMER CONNECTIONS

When a transformer has only one primary winding and one secondary winding (or one high-voltage and one low-voltage winding), it is relatively simple to make the transformer connections. About the only concern is that the primary or input is connected to the rated voltage and that the secondary or output is connected to a load that will draw no more than the rated current. It may be necessary to convert between watts and kVA, and possibly include the effects of power factor, but these are simple calculations, as described in Chapter 9.

When a transformer has multiple windings, the connections become more difficult. The same is true when two or more transformers must be used to match a given input and output voltage. In theory a transformer could have an infinite number of windings, with a corresponding number of connections. In practical work, a single-phase transformer will have no more than two high-voltage windings and two low-voltage windings. Three-phase transformers usually have no more than three high-voltage and three low-voltage windings. The following paragraphs describe these basic transformer connections.

5-2.1 Single-phase transformer connections

Figures 5-12 through 5-15 show typical single-phase transformer connections. In all four illustrations, the transformer is rated at 1.2 kVA, the two high-voltage windings are rated at 1200 V, and the two low-voltage windings are rated at 120 V.

In Fig. 5-12 the source is 2400 V, and the desired output is 120/ 240-V three-wire. Note that the tap for the neutral wire is taken at the junction of X_2 and X_3. Since the transformer is rated at 1.2 kVA, the low-voltage windings carry 5A (1200 W/240 V = 5A), whereas the high-voltage windings carry 0.5A (1200 W/2400 V = 0.5 A).

In Fig. 5-13 the source is also 2400 V, but the desired output is 120-V two-wire. Note that both low-voltage windings are connected in parallel. It is possible to obtain 120 V from only one low-voltage

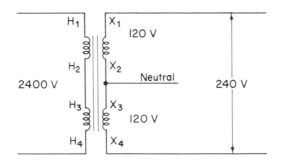

FIGURE 5-12   Single-phase   transformer   connections   for   120-240V 3-wire.

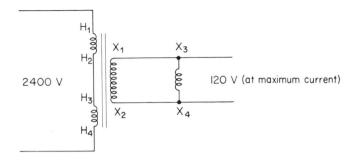

FIGURE 5-13   Single-phase transformer connections for 120V 2-wire at maximum current.

winding.   However, each low-voltage winding is capable of carrying only 5 A. If the load draws the full 10 A (to provide the full 1.2 kVA), both low-voltage windings must be used.

In Fig. 5-14 the source is reduced to 1200 V but the desired output remains at 120/240-V three-wire. The tap for the neutral wire is still taken from the junction of X_2 and X_3. However, the high-voltage windings are connected in parallel. With a 240-V output and 1.2 kVA rating, the low-voltage output windings carry 5 A. Since there is a 5:1 voltage ratio (1200/240), the current ratio is also 5:1, and the high-voltage windings must carry a total of 1 A. This 1 A is divided between the two high-voltage windings in parallel (0.5 A in each winding).

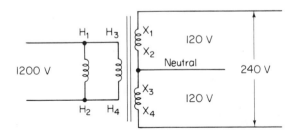

FIGURE 5-14   Single-phase   transformer   connections   for   120-240V 3-wire (with reduced high voltage).

In Fig. 5-15 the source is at 1200 V and the output is 120-V two-wire. This requires that both the high- and low-voltage windings be connected in parallel to handle the full 1.2-kVA rating.

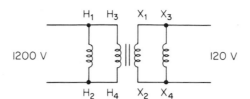

FIGURE **5-15**   Single-phase transformer connections for 120V 2-wire maximum current, and reduced high voltage.

Note: When connecting any transformer windings in parallel, always connect the odd-numbered terminals together and the even-numbered terminals together.

5-2.2 Three-phase transformer connections

Figures 5-16, 5-17, and 5-18 show typical three-phase transformer connections. The circuits apply to a three-phase transformer with

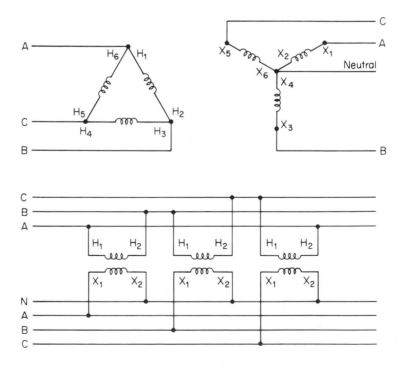

FIGURE **5-16**   3-phase transformer connections for delta-to-wye.

multiple windings or to three separate single-phase transformers. Three transformers are often used in three-phase electrical systems. One advantage is reduced transformer size. Another advantage is that a failure in one section of a multiwinding transformer usually requires replacement of the entire transformer.

The three connections shown in Figs. 5-16, 5-17, and 5-18 are delta to wye, delta to delta, and wye to delta, respectively. It is also possible to connect transformers wye to wye. However, wye to wye is not often used in interior electrical wiring. One problem is that a neutral wire must be used on both the input and output windings. The wye to wye connection is found most often on high-voltage transmission lines, which are the responsibility of the utility company.

In making any three-phase connections, it is absolutely necessary to assure that the correct phase or polarity is observed for *each section*

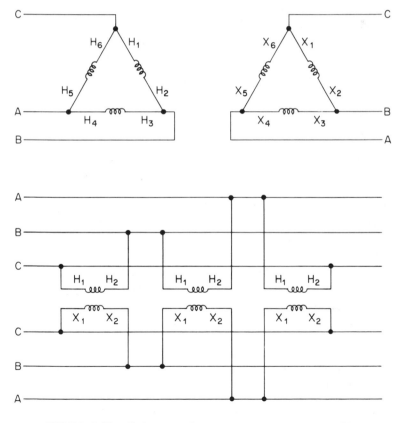

FIGURE 5-17   3-phase transformer connections for delta-to-delta.

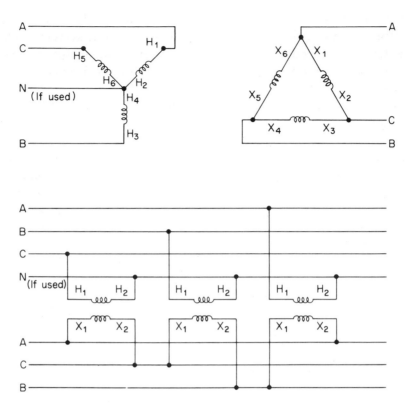

FIGURE 5-18   3-phase transformer connections for wye-to-delta.

(primary and secondary) and *from each section to all other sections.* The procedures of Section 5-1.6 can be used if there is doubt as to identification on the transformer terminals.

Note that both the geometric and line versions of the three-phase connections are shown in Figs. 5-16 through 5-18. The line versions show the terminals for three separate transformers. The geometric versions show the terminals for a single three-phase transformer with multiple windings.

5-2.3 Calculating transformer power ratings and line currents for three-phase connections

Figure 5-19 shows a typical three-phase power transformer connection. The system shown is delta to wye, using three identical trans-

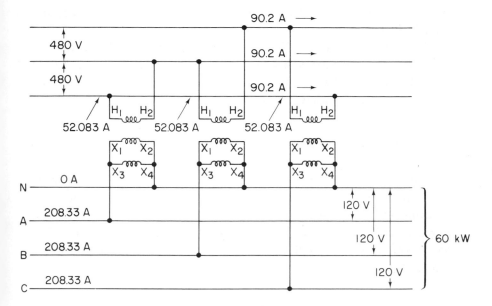

FIGURE 5-19   Calculating transformer power ratings and line currents
for 3-phase connections.

formers, each with a 480-V high-voltage winding and two 120-V low-
voltage windings. Assume that the source is 480-V three-phase and
that the load is balanced three-phase, 60 kW, with a power factor of
0.8, and requires 120 V. Find the minimum power rating (in both
watts and voltamperes) for each transformer, the current in the primary
and secondary of each transformer, the input line current from the
460-V delta source, and the output line current at the wye load.

With a 60-kW load and three transformers, the *minimum* transformer rating in watts in 60 kW/3 = 20 kW.

With a 60-kW load, three transformers, and a power factor of 0.8,
the *minimum* transformer rating in voltamperes is 60 kW/0.8 pF = 75
kVA; 75 kVA/3 = 25 kVA (per transformer).

With 25 kVA per transformer and a primary voltage of 480 V, the
primary current at each transformer is 25,000 VA/480 V = 52.083 A.

With 25 kVA per transformer and a secondary voltage of 120 V,
the secondary current at each transformer is 25,000 VA/120 V =
208.33 A.

With a total power of 60 kW, and a power factor of 0.8, the line
current in the 480-V line is 60,000/1.732 X 480 X 0.8 = 90.2 A. Or,

with a total of 75 kVA, the line current in the 480-V line is 75,000/ 1.732 X 480 = 90.2 A.

With a total power of 60 kW and a power factor of 0.8, the line current in the 120/208-V (wye-connected) line is 60,000/1.732 X 208 X 0.8 = 208.33 A. Note that this agrees with the secondary current previously calculated.

5-3 SPECIAL TRANSFORMER CONNECTIONS

In addition to conventional single-phase and three-phase transformer connections, there are many other ways in which transformers can be connected. However, there are only three systems of particular interest: autotransformer connections, the Scott or T connection, and the open delta connection. The following sections discuss each of these systems.

5-3.1 Autotransformer connections

Practically any transformer can be connected as an autotransformer. That is, the primary and secondary windings can be connected directly to produce the desired manipulation of input and output voltages. In a conventional transformer, the primary and secondary windings are electrically isolated. When the autotransformer connection is made, some of the power is conducted directly from the source to the load. In effect, the autotransformer transforms only part of the load. This makes it possible to use a transformer of a lower power rating than would normally be required.

The maximum voltages available from an autotransformer are equal to the sum of the winding voltages. For example, assume that an autotransformer has one 480-V winding (high voltage), and two 120-V low-voltage windings. The outputs available are 120 V, 240 V, 480 V, 480 V + 120 V (600 V), and 480 V + 120 V + 120 V (720 V).

The autotransformer can be operated by a source voltage equal to the voltage rating for any one winding or a combination of the winding voltages. For example, using the same autotransformer, the system could be operated from a source of 120 V, 240 V, 480 V, 600 V, or 700 V.

These voltage combinations are shown in Fig. 5-20. To use the information of Fig. 5-20, simply select the input and output terminals

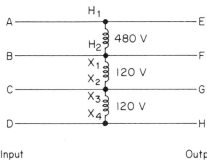

Input	Output
120 − CD	120 − GH
240 − BD	240 − FH
480 − AB	480 − EF
600 − AC	600 − EG
720 − AD	720 − EH

Example

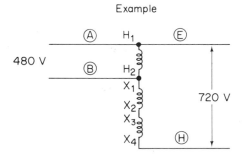

FIGURE 5-20   Basic autotransformer connections.

that match the available and desired voltages. Using the example shown
in Fig. 5-20 (input 480 V, output 720 V), the input is connected at A
and B; the output is taken from E and H. As shown, the input leads
are connected to terminals H_1 and H_2, the output leads are connected
to H_1 and X_4, and terminals H_2 to X_1, X_2 to X_3 are interconnected.

No matter what winding combination is used, the power rating of
the transformer can be reduced from the full rating by an amount
equal to the ratio of input and output voltages. For example, if the
input voltage is 480 V and the output is 720 V (as shown in Fig. 5-20),
the ratio is 2:3, and the rating can be reduced by two-thirds. Assuming
a load of 60 kVA, the transformer rating can be reduced to 20 kVA
(60 kVA X 2/3 = 40 kVA; 60 kVA − 40 kVA = 20 kVA).

This reduction of power rating can be proved using the calcula-

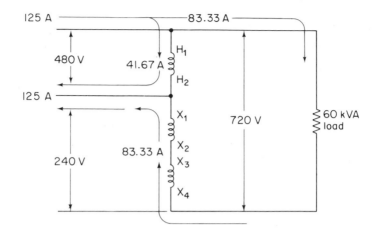

FIGURE 5-21   Example of how an autotransformer can be operated at reduced ratings.

tions of Fig. 5-21. The output (or load) current is 60,000 VA/720 V = 83.33 A. The input (or source) current is 60,000 VA/480 V = 125 A. The current through the high-voltage winding (H_1-H_2) is 125A – 83.33A = 41.67A. The kVA rating for the high-voltage winding is 41.67A X 480 V = approximately 20 kVA. The current through the low-voltage windings (X_1-X_2, X_3-X_4) is 83.33A. The kVA ratings for the low-voltage windings is 83.33A X 240 V = approximately 20 kVA.

5-3.1.1 120-V single-phase from 240-V three-phase with an auto-transformer. One of the most common uses for an autotransformer is to tap 120-V single-phase power for lighting circuits from 240-V three-phase (used for motors or similar heavy loads). Figure 5-22 shows such a system. The 240-V supply is connected as three-phase delta. The transformer is a center-tapped 240-V winding, typical of those used in 120/240-V three-wire systems. The primary winding is unused. The transformer could be a 120 V-to-120 V isolation-type transformer.

No matter what type of transformer is used, the transformer power rating can be reduced to one-half that of the load. For example, assume a 30-kVA load on the 120-V output. The output current is 30,000 VA/120 V = 250 A. The input current is 30,000 VA/240 V = 125 A. Thus the current through either winding is 125 A. The kVA rating for either 120-V winding is 125 A X 120 V = 15 kVA (or one-half of the 30-kVA load).

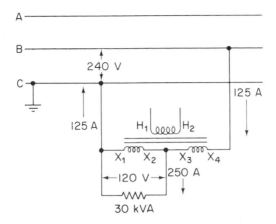

FIGURE 5-22   Example of autotransformer used to tap 120V single-phase power from 240V 3-phase.

Because of the difference in input and output currents, the transformer should be physically located as near to the 120-V single-phase load as possible. This will keep the 250-A lines at a minimum. Of course, the three-phase lines to the transformer will be lengthened, but these lines are carrying one-half the current (125 A).

Also note that one line of the three-phase supply must be grounded. *The NEC forbids use of autotransformers for lighting and appliance branch circuits unless there is a grounded conductor common to both the primary and secondary circuits.*

5-3.1.2 600-V three-phase from 480-V three-phase with an autotransformer. Another common use for autotransformers is to provide 600-V three-phase from 480-V three-phase. Electric motors typically operate from 240-, 480-, and 600-V three-phase sources. Many motors have dual windings and can operate from either 240 V or 480 V. However, 600-V heavy-duty motors usually have only one set of windings, and require the full 600 V. It is possible to use the autotransformer principle to convert 480 V to 600 V. Such a system is shown in Fig. 5-23. Three identical transformers are required. The high-voltage windings are 480 V; the low-voltage windings are 92.5 V. The output voltage is the vector sum of the two winding voltages (600 V) between each line. The system is connected as delta at both input and output.

As a guideline, the combined power ratings for all three transformers should be approximately 25 per cent of the load. For ex-

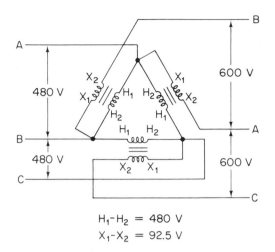

$$H_1 - H_2 = 480 \text{ V}$$
$$X_1 - X_2 = 92.5 \text{ V}$$

FIGURE 5-23   Example of autotransformer used to provide 600V 3-phase from 480V 3-phase.

ample, with an assumed load of 60 kVA, the power rating of each transformer should be 5 kVA (60 kVA X 0.25 = 15 kVA; 15 kVA/3 = 5 kVA per transformer).

5-3.2 Scott or T (two-phase) transformer connection

Practically all power distribution in the United States is either single-phase (typically three-wire 240 V) or three-phase. All electrical systems are now designed for single-phase or three-phase. However, two-phase systems still exist in certain isolated areas. In those cases where it is not practical (or economical) to change from a two-phase system, it is necessary to convert the three-phase distribution to two-phase. Figure 5-24 shows such a system.

In Fig. 5-24 the input is conventional 480-V three-phase, and the output is 240-V two-phase. This output should not be confused with conventional 240-V three-wire or with the "open delta" system described in Section 5-3.3. The output of Fig. 5-24 is a pair of 240-V voltages, displaced in phase by 90 degrees.

Generally, two transformers are used (although a special Scott transformer could be constructed). Either way, the primary windings must be tapped. One winding is center-tapped. The other winding is tapped at 0.86 of the winding voltage. With a 480-V winding the tap is at 415 V. The transformer with the center-tapped winding is called

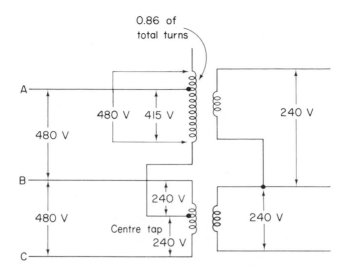

FIGURE 5-24   Example of Scott or T-transformer connection to provide 240V 2-phase from 480V 3-phase.

the *main transformer*. The transformer with the 0.86 winding tap is
called the *teaser transformer*.

The Scott transformer system can also be used to convert a two-
phase voltage to a three-phase. This is done by applying the two-
phase input at the secondaries and taking the three-phase output from
the primaries. However, such an arrangement is of little practical
value. (The Scott system was developed to save otherwise obsolete
two-phase motors, etc., when utility companies changed to single-
phase and three-phase distribution).

The Scott system has a power limitation in that the transformers
must have an approximate 116 per cent higher rating than the load.
For example, if the load is 10 kVA, the transformers must have 11.6-
kVA power ratings.

5-3.3 Open delta transformer connection

The open delta transformer connection is generally used only as
an emergency or temporary system. As shown in Fig. 5-25, the open
delta is identical to the delta-to-delta system (Fig. 5-17) except that
one transformer (or one set of primary and secondary windings) is
eliminated. If one transformer (or one set of windings) in a conventional delta-to-delta system becomes defective, simply disconnect the

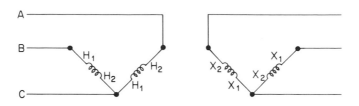

FIGURE 5-25   Open delta transformer connections.

windings (both primary and secondary) and leave the system with two transformers (or two sets of windings), as shown in Fig. 5-25.

With the open delta, the combined power rating of the two transformers is reduced to 57.7 per cent of the combined power ratings of three transformers. For example, if the three transformers in a conventional delta-to-delta system are capable of handling a 10-kVA load, the load capability will drop to 5.77 kVA. For this reason, open delta is rarely (if ever) used for original design.

5-4 INSTRUMENT TRANSFORMERS

Instrument transformers make it possible to measure high voltage and current with low-scale voltmeters and ammeters. There are two basic types of instrument transformers: potential or voltage transformers and current transformers.

5-4.1 Current transformers

Typical *current transformers* are shown in Fig. 5-26. One current transformer consists of a bar of conducting metal (typically copper or aluminum to match the system conductors) with several turns of wire wrapped around the conducting bar. In use, the conducting bar is connected in the line to be measured and a low-scale ammeter connected to the turns of wire. The ratio of turns between the two circuits depends on how large a current is anticipated and the capability of the meter. For example, with a maximum anticipated line current of 600 A and an ammeter capable of reading from 0 to 5 A (maximum), the turns ratio is 600/5 = 120:1. Now assume that 480 A passes through the line. The ammeter will indicate 4 A. In some

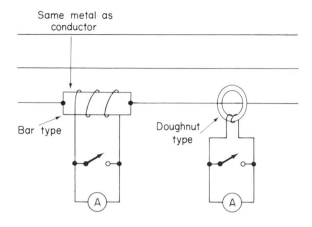

FIGURE 5-26   Typical current transformers.

cases the ammeter scale is marked with two sets of numbers (one set with a current transformer and one set without).

Another type of current transformer is the "doughnut" type. This type consists of a magnetic core in the shape of a ring. The line to be measured is passed through the ring. The ammeter is connected to windings around the ring. Operation of the doughnut type is the same as for the bar-conductor type. Current transformers are also used with the portable clamp-on type of ammeter.

Note that the current transformers shown in Fig. 5-26 are provided with a shorting switch. This switch must be closed (or a shorting strap placed across the current transformer terminals) when the ammeter is not connected. One reason for this short is that a high voltage is developed across the transformer secondary winding. For example, if the line is carrying 120 V and the turns ratio is 120:1, the winding voltage is 120 X 120 = 14,400 V. Of course, the ammeter acts as a short circuit and drops the voltage (when connected). With the ammeter disconnected, the high-voltage potential exists. An open secondary circuit can also result in a voltage drop in the primary circuit (which is carrying its own load).

From a design standpoint the main concern is that the current rating of the transformer must not be exceeded. That is, always select a current transformer with a current rating at least equal to (and preferably higher than) the maximum anticipated load. Also, make certain that the current (or turns) ratio matches that of the meter. (Or select meters to match available current transformers.)

5-4.2 Potential or voltage transformers

A typical *voltage transformer* is shown in Fig. 5-27. As shown, the voltage transformer has a high-voltage primary (connected across the high-voltage lines to be measured) and a low-voltage secondary (connected to a low-scale voltmeter).

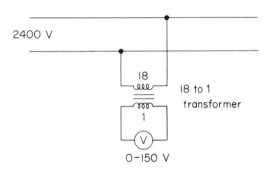

2400 V

18

18 to 1
transformer

1

0-150 V

When voltmeter is 133.33 V, line = 2400 V

FIGURE 5-27   Typical voltage transformer.

The ratio of turns depends on how large the anticipated line voltage is, and the capability of the meter. For example, assume that a 2400-V line is to be measured and the meter is capable of reading from 0 to 150 V. Further assume that the line voltage might reach 2700 V maximum. A satisfactory turns ratio is 2700/150 = 18:1. Assuming that the line remains at 2400 V, the voltmeter will read 133.33 V (unless the scale has been altered to show the actual line voltage).

As in the case of current transformers, voltage transformers are often available (in matched sets) with meters. As an alternative, the ratio data are specified on the transformer nameplate.

5-4.3 Instrument transformers for wattmeters

As discussed in Chapter 9, a wattmeter requires both voltage and current to indicate the power consumed in a system. Conventional wattmeters typically operate on voltage in the range of 120 V and at currents below 10 A. When measuring power in a high-voltage, high-current system, it is necessary to use both current transformers and voltage transformers. Such an arrangement is shown in Fig. 5-28.

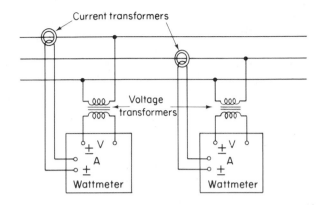

FIGURE 5-28   Instrument (voltage and current) transformers for watt-meters.

Generally, each wattmeter requires one current transformer and one voltage transformer. In the circuit of Fig. 5-28, two wattmeters are required to measure three-phase power (see Chapter 9).

5-5 TYPICAL TRANSFORMER APPLICATIONS

Since transformers can be wound with an infinite variety of turns ratios, the theoretical number of transformer applications is unlimited. However, in the practical design of interior wiring, there are only a few transformer applications. Generally, transformers are used in connection with lighting requirements, although there are exceptions (say, where a particular load requires a particular voltage and all other loads in the system require a different voltage).

As discussed in Chapter 1, most lighting systems require 120 or 277 V (also referred to as 115 and 265 V). If these voltages are not readily available at the service entrance or it is not economical to receive these voltages from the utility company, transformers are used to provide the voltages. For example, assume that an industrial building requires 480 V for electrical motors. The same building requires either 120 or 277 V (or both) for lighting. In some cases it is more economical for the utility company to supply only one voltage. In other cases the long conductors from the service entrance to the lighting, combined with the relatively low voltage (120 or 277 V compared to 480 V), will result in very high currents and large conductors. In addition, if power can be distributed from the service entrance to the lighting in three-phase form and then converted to single-phase, the current (and conductor size) can be reduced (because the power in three-phase is increased by 1.732 compared to single-phase).

In Chapter 3 we discussed the advantages (in lower currents and smaller conductor sizes) of three-phase over single-phase and higher voltage over lower voltage. These factors will not be repeated here. Instead we shall concentrate on the practical requirements for circuit design.

5-5.1 Typical transformer connections for lighting

Figures 5-29 through 5-33 show the most common arrangements for lighting transformers. In Fig. 5-29 a three-phase delta is transformed into three separate sources of 120/240-V three-wire power for distribution to lighting.

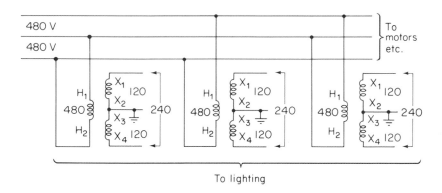

FIGURE 5-29   Connections for transforming 480V, 3-phase delta into three separate 120-240V 3-wire sources.

In Fig. 5-30, two transformer primaries are connected to the three-phase supply in an arrangement similar to the open delta (Section 5-3.3). This results in two sources of 120/240-V three-wire power for lighting.

In Fig. 5-31 a high-voltage source is converted to a single source of 120/240-V power. However, the power is distributed to two separate loads.

In Fig. 5-32 a three-phase 480-V source is transformed into a 120/208-V three-phase four-wire system. In this circuit 120 V (line to neutral) is available for lighting, with 208 V (line to line) available

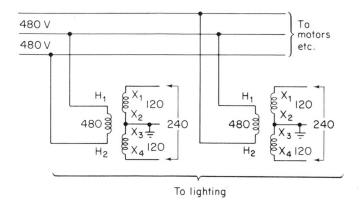

FIGURE 5-30   Connections for transforming 480V, 3-phase delta into
two separate 120-240V 3-wire sources.

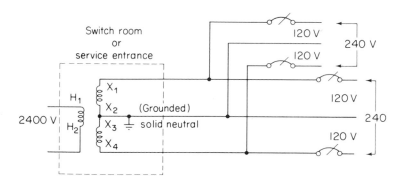

FIGURE 5-31   Connections for transforming 2400V into 120-240V
3-wire power, distributed to two separate loads.

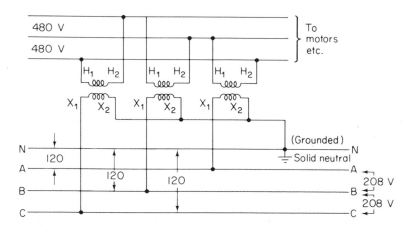

FIGURE 5-32   Connections for transforming 480V, 3-phase delta into
120-208V 3-phase, 4-wire sources.

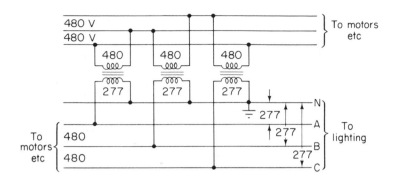

FIGURE 5-33   Connections for transforming 480V, 3-phase delta into 277-480V 3-phase, 4-wire wye distribution.

for other loads. A neutral conductor is also available for unbalanced loads.

Figure 5-33 shows a three-phase delta system transformed into a three-phase wye four-wire system at 277/480 V. This system is common for lighting in large industrial buildings, offices, and the like.

5-5.2 Design notes for lighting transformer systems

The following notes investigate the capabilities and limitations of various lighting transformer systems.

Grounding. No matter what system is used, the NEC *requires a ground or neutral conductor for all lighting.* A ground is required for the safety reasons discussed in Chapter 4. A neutral is required where there is a possibility of unbalanced loads. Any electrical system should be designed so that the loads are balanced. That is, each phase of three-phase (or both sides of 120/240-V three-wire) should have an equal load (as near as practical). Thus, if there are 100 lamps in a lighting system, 33 lamps should be connected to each phase of three-phase distribution (or 50 lamps should be connected to each side of a 120/240-V three-wire system). Under these conditions, the ground or neutral conductor current will be zero (or nearly zero). However, there is no guarantee that all lamps will be used simultaneously (except in special cases). Thus the loads can be unbalanced even in a well-designed electrical wiring system.

Placement of transformers. There are two schools of thought on transformer placement. One system of power distribution involves the installation of high-voltage three-phase or single-phase feeders throughout the building. The 120-V lighting loads (and any other loads requiring 120 V) are supplied by *transformers located near the load*, as shown in the distribution system of Fig. 5-29. With this system the load is conducted over longer distances by smaller conductors carrying smaller currents. Some manufacturers of electrical equipment provide a transformer and panelboard assembly complete with connecting ducts that can be secured to columns in industrial buildings. When using the system of Fig. 5-29, *conductors of the same size must be used to the last transformer*, and all conductors must be protected by overcurrent devices of the same ampacity.

The distribution system of Fig. 5-30 can also be arranged with the transformers near the loads. The main disadvantage of the open delta system of Fig. 5-30 is the reduction of power. The advantage is that two transformers are required instead of three.

An alternative to the systems of Figs. 5-29 and 5-30 is a system in which all feeders originate in a switch room or at the service entrance, as shown in Fig. 5-31. The utility company provides separate services for both the power and lighting, or the customer can maintain his own transformers in the area. In a large building this alternative system can result in long feeders and excessive voltage drops in the 120/240-V system. The alternative system is acceptable in smaller buildings.

Separate services. If the load is less than about 50 kW, it is generally practical for the utility company to supply two (or more) separate services (three-phase for high power and 120/240-V three-wire for lighting and small loads). When the total power requirements are greater than 50 kW, the utility company can usually furnish power at a lower rate per kilowatthour if only one type of service is provided through one meter.

Three-phase 120-V lighting. The distribution system of Fig. 5-32 can be used in place of the usual three-wire 120/240-V system in some buildings (typically apartments). The service equipment is smaller with the three-phase four-wire system shown in Fig. 5-32. If this is the only service connection to the building, the utility company usually provides the transformers (outside the building).

With the system of Fig. 5-32 it is not necessary to distribute all three phases to each apartment. Instead, three lines (neutral and two conductors) are distributed to each apartment, usually through a separate meter. The lighting and small appliances are operated from the

120 V. Large appliances must operate from 208 V. This is the main disadvantage of the 120/208-V system. All appliances are not always available in 208 V. In the case of apartment buildings, many tenants will have 240-V appliances, which creates an obvious problem.

Three-phase 277/480-V lighting. The distribution system of Fig. 5-33 can be used most effectively where large electric motor loads are combined with heavy lighting requirements. This is typical of large industrial plants. Again, if this is the only service connection to the building, the utility company usually provides the transformation, and the input to the service entrance appears as three-phase four-wire 277/480 V.

The use of 277-V lighting units permits a branch circuit 2.5 times longer than a 120-V branch circuit. This is an important advantage in large buildings, particularly those with high ceilings. Another advantage of the 277/480-V system is the availability of three-phase for motors at any point throughout the system. *The NEC permits tapping conductors of at least one-third the ampacity of the larger, if the smaller conductors do not extend more than 25 ft.* For example, assume that the main conductors running throughout the system are 100 A. Then 33-A conductors can be used to tap power for electric motors (or other loads), provided that 33-A conductors do not extend over 25 ft. The smaller conductors must be suitably protected against mechanical injury (Chapter 2) and must terminate in a single set of overcurrent devices that will limit the current to the ampacity. In our example the 33-A conductors could terminate in 33-A circuit breakers that control electric motors.

The main disadvantage of 277/480-V distribution is that only large-base lamps can be used for lighting. This limits the choice of lamps. Lamp bases of any size can be used with 120 V.) Also, if 120-V lighting must be used somewhere in the building, a separate transformer (or transformers) must be provided.

The advantages of 277-V lighting over 120-V lighting are lower currents, smaller conductors, and a higher permitted voltage drop (with the same percentage of voltage drop). In practical terms this means that 277 V is better for long runs than 120 V. As a guideline, if the branch circuits must be longer than 100 ft, use 277 V for all lighting (if possible). If the branch circuit run is between 50 and 100 ft, weigh all factors (cost, convenience, etc.) carefully. Generally, if *all lighting* can be 277 V and the run is between 50 and 100 ft, use 277 V. If the branch circuit run is less than 50 ft, there is probably little advantage in the 277-V distribution. An exception to this is where there is a very heavy lighting load (such as a large arena with thousands of lights) and all (or most) of the lighting can be 277 V.

5-6 TRANSFORMERS IN LOW-VOLTAGE CONTROL SYSTEMS

Many companies manufacture low-voltage equipment consisting of transformers, relays, and momentary contact switches for control of lights (and other loads) from many locations (either local or remote). The wiring of this remote control equipment is generally simple. Typically, a three-wire or four-wire cord is extended from switch to load. The power or line-voltage wiring to the load remains unchanged in design. The remote control wiring eliminates voltage drop in the switching circuits, since the relay performs the usual switch function. For example, assume that a lighting load of 10 A is located near the service entrance but that the lights must be controlled by a switch 500 ft from the entrance. With conventional line-voltage switch wiring, the 10 A must pass through the entire 500 ft before reaching the load. This results in a large voltage drop or very large conductors or both. In low-voltage remote-control systems, only the voltage drop between the service entrance and the load need be considered.

The transformers used in the low-voltage systems are typically 24 V (at the output or secondary) and operate from either 120 or 277 V. (However, some remote systems operate at 6 or 12 V.) No matter which voltage is used, the transformers are low-energy sources. If the low-voltage systems become short-circuited, the equipment and wiring is not subject to overcurrent. Thus danger is minimized, both to equipment and people.

5-6.1 Basic low-voltage control circuit

Figures 5-34 and 5-35 show the basic circuit and components of a General Electric remote-control low-voltage relay switching system. In this system all switches are of the single-pole double-throw normally-off momentary-contact type. Therefore, as many switches as required can be connected in parallel to a single relay, which provides multipoint switching. The motor-master control unit is a sequential-type switch that provides momentary contact for each of the 25 circuits.

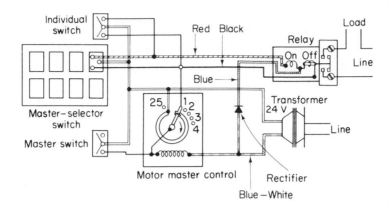

FIGURE 5-34   Basic circuit of General Electric remote-control, low-voltage relay switching system (*Courtesy* General Electric Company).

FIGURE 5-35(a)   Low-voltage remote-control switch (*Courtesy* General Electric Company).

FIGURE 5-35(b)   Low-voltage, remote-control relay (*Courtesy* General Electric Company).

FIGURE 5-35(c)   Low-voltage, remote-control transformer (*Courtesy* General Electric Company).

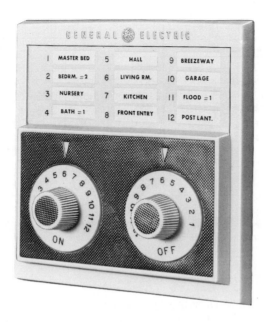

FIGURE 5-35(d)   Low-voltage, remote-control master selector switch
(*Courtesy* **General Electric Company**).

The black lead is for off control, red leads are for on control. The relay is a split-coil type, permitting positive control for on and off. When 24-V power is applied to the on-coil (red lead), the relay contacts close. When power is applied to the black lead, the relay contacts open. The relay contacts are held in either position by means of a spring. Because the actuating current flows only during the instant a switch is pressed, the relay acts as a latching-type relay and remains in the selected position. (The relay remains off until an on-current is applied, and vice versa.)

The relays can be located near the load or installed in centrally located distribution panel boxes, depending on the application. Because no power flows through the control circuits and low voltage is used for all switch and relay wiring, it is possible to place the controls at a great distance from the source or load, thus offering many advantages. For example, through the use of remote-control switching, lights for modern office buildings as well as power for equipment circuits (such as ventilating fans, air conditioners, and other service equipment) can be automated with program timers and photoelectric relays.

RR7 and RR8 (Number of relays in parallel)	Wire size (AWG)				
	12	14	16	18	20
1	9200	5800	3600	2300	1500
2	4000	2500	1600	1000	600
3	2100	1300	850	525	325
4	1100	700	450	275	175
5	550	300	200	125	75
6	150	100	75	50	25
Switch leg length run (in feet) 2 wires, using one RT1 or RT2 transformer (24 V)					

FIGURE 5-36   Typical switch runs for low-voltage, remote-control system (*Courtesy* General Electric Company).

5-6.2 Switch runs and voltage drops

Figure 5-36 shows some typical switch runs (in feet) for a low-voltage remote-control system. The values given in Fig. 5-36 apply to specific General Electric transformers and relays but are typical of low-voltage systems. The information in Fig. 5-36 is included to show a comparison of wire sizes between low-voltage remote-control systems and conventional line-voltage switching.

Assume our previous example (500 ft from switch to 10-A load), with a source of 120 V and a maximum permitted voltage drop of 6 V (5 per cent X 120 V). With conventional line-voltage switching, the conductors from switch to load would be No. 4 (for copper). No. 4 copper conductor has a resistance of 0.41 Ω for 1000 ft (500 ft between switch and load). At 10 A the voltage drop is 4.1 V. The voltage drop for No. 8 copper is 6.4 V (higher than the maximum permitted 6 V).

Using low-voltage remote-control switch wiring, it is possible to use much smaller conductors between switch and load. For example, if the load is divided between three relays, No. 18 wire can be used for the 500-ft run, as shown in Fig. 5-36.

If the load is increased, the conductors for the line-voltage switch must also be increased. However, the wire size for the low-voltage system remains the same. Similarly, since the low-voltage conductors can be smaller, more conductors can be used in a raceway of the same size. Article 725 of the NEC covers the use of remote-control, low-voltage switching. Chapter 9, Table 3 (of the NEC) lists allowable combinations of conductors in a conduit.

5-6.3 Residential low-voltage switching systems

Low-voltage switching systems are well suited to residential use. A typical wiring layout is shown in Fig. 5-37. The symbols used in Fig. 5-37 are explained in Fig. 5-38. One advantage to the home-owner is convenience. Combinations of individual switches, master switches, motor-master control units, automatic program clock controls, and photoelectric control units can be interconnected as required, providing the homeowner with flexibility not possible with conventional wiring methods. An advantage to the contractor is adaptability. The small-diameter low-voltage switch wires are easily run through thin-wall partitions, snaked in cramped areas, or concealed in modern construction that does not permit use of conventional line-voltage wires.

The layout of Fig. 5-37 shows a deluxe wiring system that employs the use of multipoint switching that permits turning on room lights from every doorway, switch-controlled split receptacles in living room and master bedroom, and a master-selector switch at the bedside. Selected lights and receptacles are also controlled from two motor-master units not shown. It has been designed to provide protection control, turning on perimeter lights to floodlight the grounds and lighting selected rooms from the master bedside. Pressing a single button S_M at any entrance, or the twelfth position of the master-selector switch, starts the motor-master for on control of the selected lights to provide a pathway of light through the house, or the off side for turning off all the lights in and around the house. In addition, a switch-controlled outdoor weatherproof split receptacle lets the home-owner turn seasonal decorative outdoor lights on and off from two handy indoor locations.

Planning a residential system. Before starting any installation it is advisable to consider the requirements of the system and the locations for such components as relay boxes and frames, motor-master control units, master-selector switches, and convenient locations for individual switches. In a residential installation the transformer is most conveniently located in one of the relay boxes, and two motor-master control units can be located in the small relay box or in one-

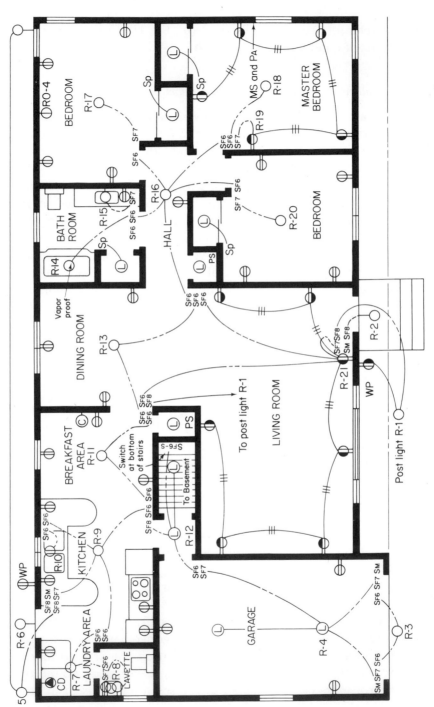

FIGURE 5-37 Typical wiring layout of General Electric low-voltage switching system (*Courtesy* General Electric Company).

158

LINE-VOLTAGE SYMBOLS	
Symbol	Description
———	2-Cond. 120 V Wire or Cable
—///—	3-Cond. 120 V Wire or Cable
O	Ceiling Receptacle
Q	Floodlight
⊏=O=⊐	Valance Light
-Ⓒ	Clock Receptacle
Ⓛ	Keyless Lampholder
Ⓛ$_{PS}$	Pull Chain Lampholder
=⊖	Double Receptacle, Split Wired
≡⊖	Grounding Receptacle
≡⊖$_{WP}$	Weatherproof Grounding Receptacle
≡⊖$_R$	Range Receptacle
-Ⓞ$_{CD}$	Clothes Dryer Receptacle
S$_L$	Lighted-Handle Mercury Switch
S$_P$	Push-Button, Pilot Switch
S$_D$	Closet Door Switch

LOW-VOLTAGE SYMBOLS	
Symbol	Description
—··—	Remote-Control Low-Voltage Wire
T	Low-Voltage Tranformer
—▶—	Rectifier for Remote Control
B$_R$	Box for Relays and Motor Master Controls
R	Remote-Control Relay
R$_P$	Remote-Control Pilot-Light Relay
P$_{11}$	Separate Pilot Light, R.C. Plate
P$_{10}$	Separate Pilot Light, Inter. Plate
M$_S$	Master-Selector Switch
MM$_R$	Motor Master Control for ON
MM$_B$	Motor Master Control for OFF
S$_M$	Switch for Motor Master
S$_{F6}$	R. C. Flush Switch
S$_{F7}$	R. C. Locator-Light Switch
S$_{F8}$	R. C. Pilot-Light Switch
S$_{K6}$	R. C. Key Switch
S$_{K7}$	R. C. Locator Light Key Switch
S$_{K8}$	R. C. Pilot-Light Key Switch
S$_{T6}$	R. C. Trigger Switch
S$_{T7}$	R. C. Locator Light Trigger Switch
S$_{T8}$	R. C. Pilot-Light Trigger Switch
S$_{T4}$	Interchangeable Trigger Switch, Brown
S$_{T5}$	Interchangeable Trigger Switch, Ivory
⊖$_{RO}$	Remote-Control Receptacle for Extension Switch

FIGURE 5-38 Symbols used in General Electric low-voltage remote-control systems (*Courtesy* General Electric Company)

half of the larger box. These relay boxes should be located close to the service entrance equipment.

For large houses it may be advisable to split up the relay boxes, locating one or two in far sections of the house to shorten the run of 120-V wiring from relays to controlled receptacle boxes.

Consider the location of the master-selector switch in the garage, close to the car door, and the location of the master bedroom, either between twin beds or at the doorway. An extension switch on the night table can be used to "master" control the motor-master units.

If a layout has not been provided, it is advisable to make a wiring layout, similar to that of Fig. 5-37, before starting the installation. Any portion of the Fig. 5-37 layout can be used as a guide in laying out a low-voltage remote-control installation to meet requirements specified by the architect, builder, or home buyer. Note that the symbols used in Figs. 5-37 and 5-38 are those suggested by General Electric for their low-voltage remote-control systems.

Chapter 6

Interior Lighting Design

In this chapter we shall discuss design of electrical lighting from the standpoint of providing a given amount of light for a given area. That is, we shall discuss calculations to determine how many lamps are required to provide a given illumination level in a given interior space and how the lamps should be arranged to provide uniform illumination throughout the space. With the number of lamps (of given wattage or voltage and current rating) established, the electrical wiring system (conductor size, voltage drop, transformer connections, etc.) for the lamps can be calculated using the information of Chapters 1 through 5.

There is considerable disagreement among lighting experts as to how much light is required for a given situation. In many cases the amount of light is a matter of preference. In any event, there are many charts and tables available from lighting equipment manufacturers and the various engineering societies which show their recommendations for the amount of light in given areas and situations. These will not be duplicated here. We shall concentrate instead on how to provide a given amount of light, in a given area, under a given set of conditions.

6-1 ELECTRICAL LIGHTING BASICS

Before going into practical design, let us review the basics of electrical lighting, such as the terms used, most common light sources, efficiency factors, and depreciation factors.

6-1.1 Lamps and luminaires

Except in special cases, most electrical lighting is provided by incandescent or fluorescent *lamps*. Lamps are held by *luminaires*. Luminaires may be functional or decorative (or both) and are usually referred to by the general public as lighting fixtures. In addition to holding the lamps, luminaires function to distribute and diffuse the light from the lamps. All luminaires have certain efficiency and depreciation factors. That is, in an effort to distribute and diffuse the light so that glare is minimized and lighting is uniform, luminaires cannot be designed for 100 per cent efficiency. (A single unshaded lamp hanging from a cord in the center of a room is nearly 100 per cent efficient, but the glare makes such an arrangement impractical.)

6-1.2 Efficiency of luminaires

Many factors affect the efficiency of luminaires. A few of the important factors are construction of the luminaire, the type of light source, reflection from surrounding areas, cleanliness of the light sources over a period of time, and the distances from the light source to the *work plane* (the point where the level of light intensity is to be measured). Distance is a very important factor. Light varies inversely as the square of the distance from the light source. For example, if the distance doubles, the light is reduced to one-fourth.

6-1.3 Coefficient of utilization

Many of these foregoing factors, and others, are combined into a single efficiency factor called the *coefficient of utilization* (CU). The CU for a particular luminaire is determined by experiment and is given by the luminaire manufacturer in its catalog. Typically the manufacturer's publications include charts that give consideration to the rela-

tive light-absorbing qualities of the ceiling, walls, floors, and the luminaire. Also shown is the relative pattern of light, illustrating what
percentage is reflected up to the ceiling or down to the floor. Such
charts are illustrated in later sections of this chapter.

6-1.4 Lumens and footcandles

Lumens and *footcandles* are the most common terms used in the
measurement of light. Some manufacturers, and some lighting specifications, also use the terms *footlamberts, candelas per square inch,
reflectance, transmittance,* and *absorbance.* All these terms are defined briefly in the following paragraphs.

The *candela* (pronounced can-*del*-la, and formerly known as
candle) is the unit of luminous intensity of a light source in a specified direction. The standard candela (cd) is defined as 1/60 the
intensity of a square centimeter of a black-body radiator operated at
the freezing point of platinum ($2047°$ K).

To measure the luminous output of lamps and luminaires, a unit
for the light-producing power of a light source, the *lumen,* came into
being. The lumen (lm) is defined as the rate at which light falls on a
1-ft^2 surface area that is equally distance (1 ft) from a source whose
intensity is 1 cd. Lamps are commonly rated by their *total lumen
output.* Also, the number of *lumens per watt* represents what is called
the efficiency of the light source.

Illumination on a surface is measured in *footcandles.* A footcandle (fc) is defined as the density of light striking each and every
point on a segment of the inside surface of an imaginary 1-ft radius
sphere with a 1-candlepower source at the center (see Fig. 6-1). Or,
1 fc is the illumination on 1 ft^2 of surface over which 1 lm is distributed evenly. This means that 1 fc equals 1 lm/ft^2. The footcandle
is the key unit in engineering calculations for lighting installations.
Where the metric system is used, the imaginary spherical surface is
given as 1 m^2, and the radius from the light source is 1 m. When 1 lm
is distributed evenly over the square meter, the illumination is said to
be 1 *lux.* The ratios among candelas, lumens, footcandles, and lux
are given in Fig. 6-1. The terms "lumen" and "footcandles" are used
throughout this chapter.

Light is invisible as it travels through space. What the eye actually
sees is brightness, the result of light being reflected or emitted by a

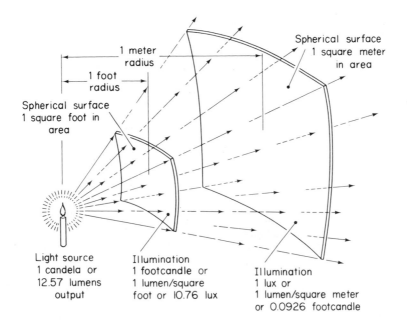

FIGURE 6-1   Relationship of lumens, footcandles, candelas, and lux
(Courtesy **General Electric Company).**

surface directly into the eye. There are two common units of brightness: *footlamberts* and *candelas per square inch.* A surface that emits or reflects light at the rate of 1 lm/ft^2 of area has a brightness of 1 footlambert (fL), as viewed from any direction. One candela per square inch equals 452 fL (see Fig. 6-2).

Reflection factor *(reflectance)*, transmission factor *(transmittance)*, and absorption factor*(absorptance)* are values that indicate the amount of the total light on the surface that is reflected, transmitted, and absorbed. For example, assume that 100 lm strikes a translucent spherical surface of 1 ft^2 as shown in Fig. 6-2. The incident illumination is said to be 100 fc. If 60 lm of light is reflected, the brightness of the surface on the reflected light side is 60 fL and the reflection factor is 0.6 or 60 per cent. If 30 lm is transmitted through and emitted, the brightness on the other side of the surface is 30 fL. The transmission factor is 0.3 or 30 per cent. Since 60 lm is reflected and 30 lm is transmitted, 10 of the initial 100 lm are absorbed. The absorption factor is thus 0.1 or 10 per cent.

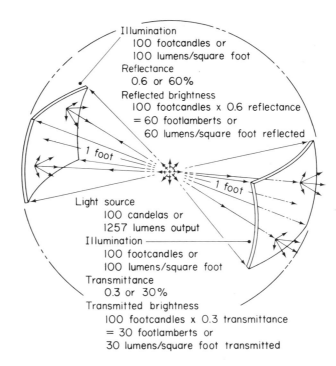

Illumination
   100 footcandles or
   100 lumens/square foot
Reflectance
   0.6 or 60%
Reflected brightness
   100 footcandles x 0.6 reflectance
   = 60 footlamberts or
      60 lumens/square foot reflected

1 foot

1 foot

Light source
   100 candelas or
   1257 lumens output
Illumination
   100 footcandles or
   100 lumens/square foot
Transmittance
   0.3 or 30%
Transmitted brightness
   100 footcandles x 0.3 transmittance
   = 30 footlamberts or
      30 lumens/square foot transmitted

FIGURE 6-2   Relationship of footlamberts, reflectance, and transmittance of light (*Courtesy* General Electric Company).

6-2 LIGHT SOURCES

As stated, the most commonly used light sources are fluorescent and
incandescent lamps. In addition, there are special lamps, such as high-
intensity discharge (mercury) and vapor lamps. Because of their
unique characteristics, we shall not discuss these special lamps, nor
shall we discuss the basic theory of incandescent or fluorescent lamps.
Instead, we shall concentrate on the lamp characteristics as they
apply to practical lighting design.

6-2.1 Fluorescent-lamp design characteristics

Fluorescent lamps have many design advantages over incandescent
lamps. Fluorescent light is distributed over a larger lamp surface area.
Fluorescent lamps produce more lumens per watt than incandescent

lamps. In properly designed luminaires, fluorescent lamps can distribute their lumen output in a manner that approaches the ideal situation (where an entire ceiling area emits light evenly). Fluorescent lamps have certain disadvantages. One of the drawbacks is that fluorescent lamps (unlike incandescents) must be suitably protected (in separate enclosures) against humidity and temperature extremes.

There are two basic types of fluorescent lamps. One type has two contact pins in each end. The other type has one contact pin at each end. Thus there are two basic types of fluorescent lamp sockets. Both types require auxiliary equipment, such as transformers, ballasts, starters, and capacitors. All these auxiliary components are usually part of the luminaire. The capacitor serves two purposes and is of particular importance in design. One purpose of the capacitor is to minimize the stroboscopic effect of fluorescent lamps. The capacitor in series with one lamp enables the current in the lamp to be out of phase with that of the other lamp. This causes the lamps to go on and off at different times with respect to the ac input voltage. If all fluorescent lamps in a luminaire operated together, without phase shift, a stroboscopic effect would be noted on moving objects. (This is no problem with incandescent lamps since their filaments continue to glow when alternating current passes through zero.)

The other purpose of the capacitor is to increase the power factor. The ballasts used with fluorescent lamps are highly inductive and without a capacitor will produce a power factor of about 0.6. With the capacitor the power factor is increased to about 0.9. Fluorescent lamps are not rated by voltage. Instead, the transformers and ballasts are designed for specific voltages. The most frequently used voltages for fluorescent luminaires are 120, 208, 240, and 277 V (known by some manufacturers as 115, 200, 230, and 265V). *The NEC requires that all fluorescent luminaires be plainly marked with their voltage and current ratings, including ballasts and transformers.* The frequency must also be included.

Fluorescent lighting requires more consideration when current and power are being calculated. For one reason, the NEC does not require that the power rating of the luminaire be given. Thus current, rather than power, determines the size of the distribution system (conductor size, voltage drop, etc.). Another reason is that the stated current rating is lower for fluorescent luminaires with power factor correction, than those without power factor correction. Luminaires with power factor correction (capacitor) are usually identified as HPF (high-power-factor) units.

If the wattage of the fluorescent lamps is known but the current for the luminaire is not known, the approximate current can be found as follows. Add the wattage of all lamps in the luminaire. Add 20 per cent of the wattage to the total. Multiply the luminaire voltage by 0.9 (for a HPF unit) or by 0.6 (for an uncorrected unit). Divide the total wattage by this factor. For example, assume that four 80-W lamps are used in a fluorescent luminaire and that the voltage is 120 V. Find the current rating for both uncorrected and HPF luminaires: 80 W X 4 = 320 W; 320 W X 0.20 = 64 W; 320 W + 64 W = 384 W; 384 W/120 X 0.9 = approximately 3.58 A for HPF units; 384 W/120 X 0.6 = approximately 5.33 A. It is obvious that the capacitor-corrected HPF units are far more efficient.

Lamp		Lumens after 100 hours (initial)	Length (including sockets) (inches)
20	watt, T−12		24
	daylight	910	
	cool white	1081	
	white and warm white	1110	
40	watt, T−12		48
	daylight	2350	
	cool white	2900	
	soft white	1900	
	white and warm white	3000	
85	watt, T−12		72
	cool white	5550	
105	watt, T−12		96
	warm white	7800	
	cool white	7900	

FIGURE 6-3   Typical fluorescent lamp sizes and lumen ratings.

Figure 6-3 shows some typical fluorescent lamp sizes. Note that diameters of the lamps are given in eighths of an inch. For example, T-12 is a diameter of 12/8, or $1\frac{1}{2}$, in.

Figure 6-4 shows sizes and characteristics of typical *slimline* fluorescent lamp. Slimline lamps have only one contact at each end.

The life of fluorescent lamps is about 7500 hours (h) for average use. The lumen output drops with age. Typically, the lumen output will drop to about 80 per cent after 6000 to 7000 h of operation.

The life of fluorescent lamps can also be estimated by the number of starts. Three hours per start is a good rule of thumb. That is, if

Size	Length (inches)	Watts	Current (mA)	Lumens after 100 hours (initial)
T−8	72	34	120	1770
T−8	72	49	200	2600
T−8	72	56	300	3100
T−12	48	23	200	1600
T−12	48	39	425	2700
T−12	96	45	200	3400
T−12	96	74	425	5700

FIGURE 6-4   Typical slimline fluorescent lamp sizes, currents, and lumen ratings.

the normal lamp age is 7500 h, this assumes no more than about 2500 starts. If the number of starts is increased on the same lamp, the life will be shortened in proportion.

Rapid-start fluorescent lamps are available. These lamps do not have conventional starters but do require auxiliary equipment. For example, the auxiliary equipment in slimline lamps causes the lamp to start in about 2 seconds (s).

Fluorescent lamps will operate satisfactorily within a range of about 6 per cent of the rated voltage. This is shown in Figs. 6-5 and 6-6. For example, if the voltage is increased to about 6 per cent above the normal rated voltage, the wattage will increase to about 108 per cent of normal (Fig. 6-5), whereas the lumens output will increase to

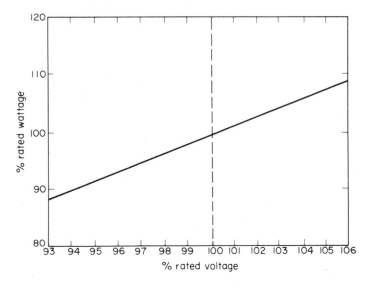

FIGURE 6-5   Typical wattage of fluorescent lamps, versus voltage.

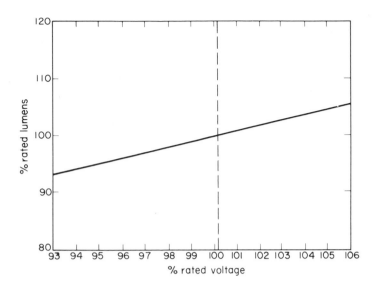

FIGURE 6-6   Typical lumen output of fluorescent lamps, versus voltage.

about 106 per cent (Fig. 6-6). Decreases in operating voltage cause corresponding decreases in both wattage and lumens. This factor must be considered in design of electrical distribution systems, particularly with regard to voltage drop in conductors to the lighting fixture. *Any system that produces a voltage drop greater than about 5 per cent of the normal rated voltage for the lamp (or luminaire) is not satisfactory.*

6-2.2 Incandescent-lamp design characteristics

Incandescent lamps use a tungsten filament. The resistance of the cold tungsten filament is very low in relation to the resistance when the filament is heated. The low resistance produces a very large current flow when the lamp is first turned on. This high current causes the lamp to glow quickly (typically in about 0.1 s). As far as the human eye is concerned, the incandescent lamp goes on instantly. Once the lamp is on and the filament is heated, the resistance increases rapidly to reduce the current flow.

For practical design, the *high inrush current* characteristic of incandescent lamps is of particular importance in the selection of switches (and relay contacts of the low-voltage switching systems described in Chapter 5). The contacts of switches and relays used in incandescent lighting systems must be capable of handling high currents even though the currents are momentary. Such switches are called *T-rated switches*. The current rating of T-rated switches is for the inrush current of the lamp, not the continuous current that is drawn after the lamp is on. For example, a 100-W incandescent lamp requires 0.833 A of continuous operating current at 120 V. However, the inrush current of the lamp is about 8.5 to 9 A. The switch for such a lamp should have a 10-A T rating (or higher).

Figure 6-7 shows some typical incandescent lamp sizes, together with outputs in lumens and average life expectancy. Note that both *initial lumens* and *average lumens* are given. The initial lumen figures are those of new lamps just installed. The average lumens are those that can be reasonably expected over the rated life. Both factors are used in lighting design, as discussed in later sections of this chapter.

	Standard base lamps		
Watts	Initial lumens	Average lumens	Life (in hours)
25	265	230	1000
40	470	415	1000
60	840	795	1000
75	1230	1100	1000
100	1640	1540	750
150	2700	2420	750
200	3800	3400	750
300	5750	5150	1000
	Mogul base lamp		
300	5750	5150	1000
500	10,000	8700	1000
750	16,500	14,700	1000
1000	23,000	20,000	1000

FIGURE 6-7   Typical incandescent lamp ratings.

Incandescent lamps have certain advantages over fluorescent lamps. Incandescent lamps do not require the same degree of protection against humidity and temperature extremes. Incandescent lamps can be operated in practically any environment. No auxiliary equipment is required for incandescent lamps. Another advantage is that the load presented by incandescent lamps is purely resistive. This eliminates the need for power factor correction, required with fluorescent lamps. It also makes power and current calculations for incandescent lamps relatively simple (current is found when wattage rating is divided by voltage).

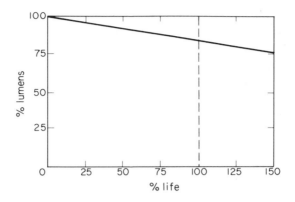

FIGURE 6-8   Light loss (drop in lumen output) over rated life of a typical incandescent lamp.

All lamps deteriorate with age. That is, the lumen output continues to drop from the initial value until the lamp finally burns out. Figure 6-8 shows the light loss (drop in lumen output) over the rated life of a typical incandescent lamp. When the lamp has been used its full rated life (100 per cent), the lumen output is about 88 per cent of the initial value. If the lamp should last 150 per cent of its rated life, the lumen output drops to about 70 per cent of the initial value.

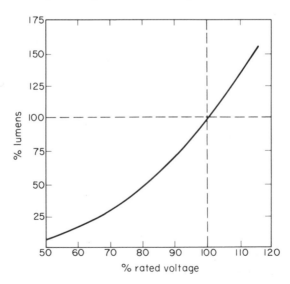

FIGURE 6-9   Typical lumen output of incandescent lamps, versus voltage.

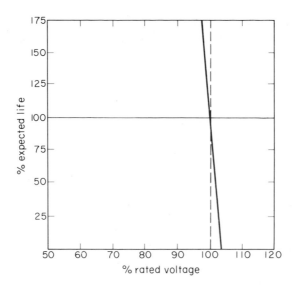

FIGURE 6-10   Typical change in expected life of incandescent lamps, versus voltage.

Note that even though the lumen output drops, the wattage consumption remains almost constant over the life expectancy of the incandescent lamp. There is about a 4 or 5 per cent drop in wattage consumption, compared to a 12 or 13 per cent drop in lumen output, at the full life expectancy. Relamping should be done regularly, based on the value of Fig. 6-8, if it is desired to maintain a given light level.

The effects of operating incandescent lamps at voltages other than their rated voltage is more pronounced than for fluorescent lamps. This is shown in Figs. 6-9 and 6-10. For example, a 5 per cent drop in rated voltage will reduce the rated lumen output to about 80 per cent. (A fluorescent lamp will drop the lumen output to about 95 per cent for a 5 per cent drop in voltage, as shown in Fig. 6-6.) On the other hand, an increase in operating voltage of about 5 per cent (above the normal rated voltage) will decrease the life expectancy of incandescent lamps to less than 50 per cent of the rated value.

6-3 ZONAL CAVITY METHOD OF LIGHTING DESIGN

This section is written to illustrate a simple, practical approach to the problem of designing interior general lighting systems, use of the *zonal cavity method.* The technique described has been developed by the General Electric Company using standard Illuminating Engineering Society (IES) tables as the basis. However, the procedure represents a modification of the conventional IES methods.

Two typical example problems are explained in detail to indicate the flexibility of the zonal cavity method; shortcuts and rules of thumb are included for those instances when a fast, approximate answer will suffice.

The zonal cavity method is preferred for general lighting calculations not because it is necessarily more accurate, but because it is more flexible, can be applied to every type of interior space, and allows more of the actual lighting situation to be taken into account during the calculations. Numerical results are thus generally more representative of the actual situations than those arrived at by older calculation systems (using the same IES tables). This is especially true when odd sizes and shapes of rooms are involved or when surface mounted, recessed, or luminous ceiling lighting systems are used.

6-3.1 Calculating illumination

Interior lighting calculations are usually used to determine

1. How many luminaires are required to provide an average given illumination level in an interior space.
2. How the luminaires should be arranged to provide uniform illumination throughout the space.

Conversely, if the quantity and type of equipment is known for a given space, the illumination level may then be calculated.

Illumination calculations are based on the fact that

$$\text{illumination (footcandles)} = \frac{\text{lumens}}{\text{area (in square feet)}}$$

or

$$\text{fc} = \frac{\text{lumens}}{\text{ft}^2}$$

which means that illumination is the result of lumens generated, divided by the lighted area.

Of course, this is for an ideal case, which does not take into account absorbtion of light by wall, ceiling, and floor surfaces; the interreflection of light; the efficiency or distribution of the luminaire; the shape of the room; or even where the illumination is measured (floor, desktop, counter, etc.).

By reducing all these factors to a *single quantity*, the ideal case can be made into a realistic case. The quantity in most common use by lighting engineers and the IES is the *coefficient of utilization* (CU), discussed in Section 6-1.3. Taking the CU into account, the basic equation becomes

$$\text{illumination (footcandles)} = \frac{\text{lumens}}{\text{area (in square feet)}}$$

$$\text{X (coefficient of utilization)}$$

or

$$\text{fc} = \frac{\text{lumens}}{\text{ft}^2} \text{ X CU}$$

The zonal cavity method of lighting design is a systematic way of determining an accurate CU, using conventional IES tables.

To account for the depreciation of illumination over time due to dirt on room and luminaire surfaces and aging of the light sources, two other factors must be included in the equation: (1) *lamp lumen depreciation* (LLD) and, *luminaire dirt depreciation* (LDD).

The equation is now complete and represents *average maintained illumination* resulting from a given luminaire in a given room:

$$\text{fc} = \frac{\text{lumens}}{\text{ft}^2} \text{ X CU X LLD X LDD}$$

An equation to determine the number of luminaires required to provide a given average maintained illumination level can be developed by a simple rewriting of the equation

$$\text{number of luminaires} = \frac{\text{maintained footcandles X area}}{\text{CU X (lumens per lamp) X (lamps per luminaire) X LDD X LLD}}$$

An alternative equation that provides luminaire spacing information can be written

$$\text{area/luminaire} = \frac{\text{(lumens per lamp) X (lamps per luminaire) X CU X LDD X LLD}}{\text{maintained footcandles}}$$

Then the total number of luminaires required is

$$\text{total luminaires} = \frac{\text{total room area}}{\text{area/luminaire}}$$

6-3.2 Calculating the basic equations

The quantities needed to solve the equation can be found by use of the following step-by-step procedures:

1. Select the lamp and luminaire. Use either the manufacturer's data or the standard IES data. Samples of such data are included in Fig. 6-11. The information shown is presented in the standard IES format.   Manufacturers do not always follow this format exactly but usually produce similar formats in their publications. Some manufacturers make reference to the IES. For example, the publications of some manufacturers simply specify that their particular lamp or luminaire has the characteristics of IES luminaire 33.

Although the exact format is not critical, certain lamp and luminaire information is necessary for the zonal cavity method. Such information includes the maximum spacing ratio in relation to mounting height above the work plane (shown as Max. S/MH$_{W\rho}$ in Fig. 6-11), the room cavity ratio (shown as RCR), the per cent of effective ceiling cavity reflectance (shown as ρ_{CC}), per cent wall reflectance (ρ_W), maintenance category (LDD Maint), and the CU for 20 per cent effective floor cavity reflectance (ρ_{FC}). A drawing or sketch of the luminaire, together with a brief description of its characteristics (dimensions, angle of light distribution, percentage of upward and downward light distribution, etc.), is also helpful.

2. Decide on the *average maintained* illumination required. Use manufacturers' recommendations, architects' requirements, IES standards, or customer preference.
3. Calculate the work plane area. Record the total length and total width of the work plane.
4. Calculate the CU, as described in Section 6-3.3.
5. Look up the lumens per lamp (in the lamp manufacturer's catalog).
6. Look up the LLD, which is usually given in manufacturers'

Typical Distribution and Maximum Spacing	ρcc →	80			70			50			30			10			0	Typical Luminaires and Luminaire Maintenance Category
	ρw →	50	30	10	50	30	10	50	30	10	50	30	10	50	30	10	0	
	RCR ↓	Coefficients of Utilization for 20 Per Cent Effective Floor Cavity Reflective, ρ_{FC}																
21 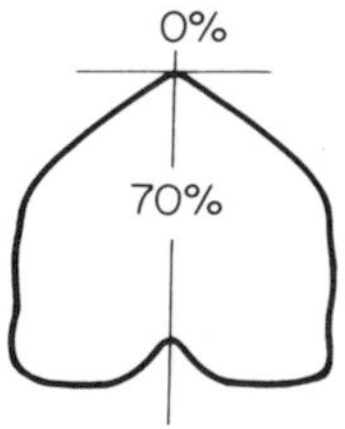	1	.78	.77	.74	.76	.75	.73	.74	.72	.71	.71	.70	.68	.68	.67	.66	.65	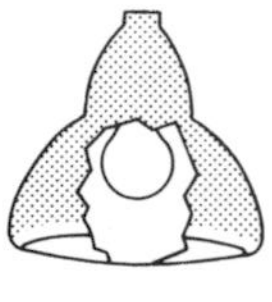 Enclosed reflector with incandescent lamp LDD Maint. Category V
	2	.72	.68	.66	.71	.67	.65	.68	.66	.63	.65	.64	.62	.64	.62	.61	.59	
	3	.66	.62	.59	.65	.61	.58	.63	.60	.57	.61	.59	.56	.60	.57	.55	.54	
	4	.60	.56	.52	.59	.55	.52	.58	.54	.51	.56	.53	.51	.55	.53	.50	.49	
	5	.55	.50	.47	.54	.50	.46	.53	.49	.46	.52	.48	.46	.51	.48	.45	.44	
	6	.51	.45	.42	.50	.45	.42	.49	.45	.41	.48	.44	.41	.47	.43	.41	.40	
	7	.46	.41	.37	.46	.41	.37	.45	.40	.37	.44	.40	.37	.43	.39	.36	.35	
	8	.42	.37	.33	.42	.37	.33	.41	.36	.33	.40	.36	.33	.39	.35	.33	.31	
	9	.39	.33	.30	.38	.33	.30	.37	.33	.30	.37	.32	.29	.36	.32	.29	.28	
Max. S/MH$_{wp}$ = 1.5	10	.33	.28	.25	.33	.28	.25	.32	.28	.25	.32	.28	.24	.31	.27	.24	.23	
22 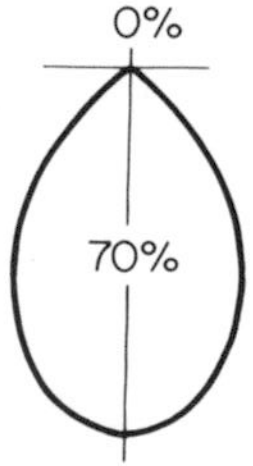	1	.75	.73	.71	.74	.72	.70	.71	.69	.68	.68	.67	.66	.66	.65	.64	.62	 Medium distribution ventilated aluminum or glass reflector with improved-color mercury lamp LDD Maint. Category III
	2	.68	.64	.61	.67	.63	.61	.64	.62	.59	.62	.60	.58	.60	.58	.57	.55	
	3	.62	.57	.54	.60	.56	.53	.59	.55	.52	.57	.54	.51	.55	.53	.50	.49	
	4	.56	.52	.47	.55	.51	.47	.53	.49	.46	.52	.48	.45	.50	.47	.45	.44	
	5	.50	.45	.41	.49	.44	.41	.48	.44	.40	.47	.43	.40	.46	.42	.40	.39	
	6	.45	.40	.36	.45	.40	.36	.43	.39	.36	.43	.38	.35	.41	.38	.35	.34	
	7	.41	.36	.32	.40	.35	.32	.40	.35	.31	.38	.34	.31	.38	.34	.31	.30	
	8	.37	.32	.28	.37	.32	.28	.36	.31	.28	.35	.31	.28	.34	.30	.28	.26	
	9	.33	.28	.24	.33	.28	.24	.32	.28	.24	.32	.27	.24	.31	.27	.24	.23	
Max. S/MH$_{wp}$ = 1.2	10	.30	.25	.22	.30	.25	.22	.29	.25	.22	.29	.25	.22	.28	.24	.21	.20	

Ratio of maximum spacing between luminaire centers to mounting (or ceiling) height above the work plane.
RCR = Room Cavity Ratio.
ρ_{cc} = Per cent effective ceiling cavity reflectance.
ρ_w = Per cent wall reflectance.

FIGURE 6-11   Coefficients of utilization for luminaire types 21 through 24, per IES standards. (Part 1)

Typical Distribution and Maximum Spacing	ρ_{CC} → ρ_W → RCR ↓	80			70			50			30			10			0	Typical Luminaires and Luminaire Maintenance Category
		50	30	10	50	30	10	50	30	10	50	30	10	50	30	10	0 ρ_{FC}	
		Coefficients of Utilization for 20 Per Cent Effective Floor Cavity Reflective, ρ_{FC}																
23 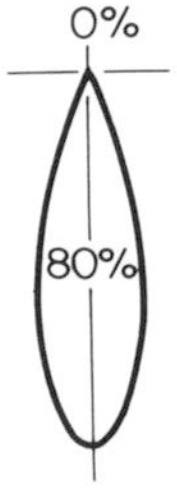 0% 80% Max. S/MH$_{wp}$ = 0.6	1	.89	.87	.85	.87	.85	.84	.84	.82	.81	.81	.80	.79	.78	.77	.77	.75	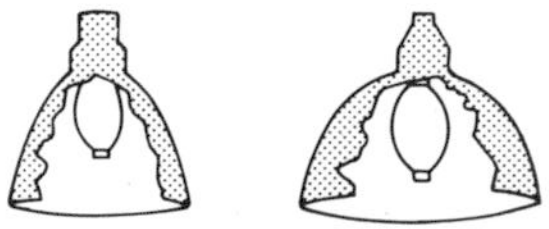 400 W 1000 W Narrow distribution, ventilated aluminum or glass reflector with clear mercury lamp LDD Maint. Category III
	2	.82	.79	.76	.81	.78	.76	.78	.76	.74	.76	.74	.72	.74	.72	.71	.69	
	3	.76	.72	.69	.75	.71	.69	.73	.70	.67	.71	.69	.66	.69	.67	.65	.64	
	4	.71	.66	.63	.70	.66	.62	.68	.65	.62	.67	.64	.61	.65	.62	.60	.59	
	5	.66	.61	.57	.65	.60	.57	.63	.59	.56	.62	.59	.56	.61	.58	.55	.54	
	6	.61	.56	.53	.61	.56	.53	.60	.55	.52	.59	.55	.52	.57	.54	.52	.50	
	7	.57	.52	.48	.56	.52	.47	.55	.51	.48	.54	.50	.48	.54	.50	.47	.46	
	8	.53	.48	.44	.52	.47	.44	.51	.47	.44	.51	.47	.44	.50	.46	.43	.42	
	9	.49	.43	.40	.48	.43	.40	.47	.43	.40	.47	.42	.40	.46	.42	.39	.38	
	10	.45	.40	.37	.45	.40	.37	.44	.39	.36	.43	.39	.36	.43	.39	.36	.36	
24 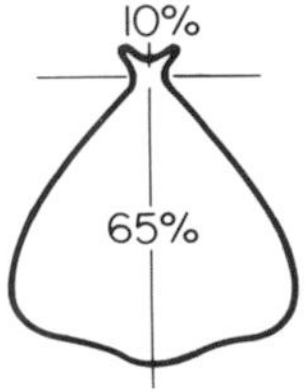 10% 65% Max. S/MH$_{wp}$ = 1.4	1	.81	.79	.77	.77	.76	.74	.73	.71	.70	.68	.67	.66	.63	.63	.62	.60	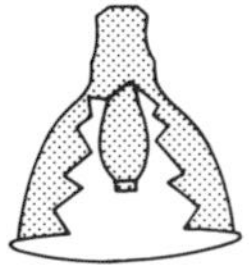 Wide distribution, ventilated aluminum or glass reflector with improved-color mercury lamp LDD Maint. Category III
	2	.74	.71	.68	.72	.69	.66	.67	.65	.63	.63	.62	.60	.60	.58	.57	.55	
	3	.68	.64	.61	.66	.63	.59	.63	.60	.57	.59	.57	.55	.56	.54	.52	.51	
	4	.63	.58	.54	.61	.57	.53	.58	.54	.51	.55	.52	.49	.52	.50	.48	.46	
	5	.57	.52	.49	.56	.51	.48	.53	.49	.47	.51	.47	.45	.48	.46	.43	.42	
	6	.53	.48	.44	.52	.47	.43	.49	.45	.42	.47	.43	.41	.45	.42	.40	.38	
	7	.48	.43	.40	.47	.42	.39	.45	.41	.38	.43	.40	.37	.41	.38	.36	.35	
	8	.44	.39	.36	.43	.38	.35	.41	.37	.34	.40	.36	.33	.38	.35	.32	.31	
	9	.41	.35	.32	.40	.35	.31	.38	.34	.31	.36	.33	.30	.35	.32	.29	.29	
	10	.35	.30	.27	.35	.30	.26	.33	.29	.26	.32	.28	.25	.30	.27	.24	.23	

Ratio of maximum spacing between luminaire centers to mounting (or ceiling) height above the work plane.

RCR = Room Cavity Ratio.

ρ_{CC} = Per cent effective ceiling cavity reflectance.

ρ_W = Per cent wall reflectance.

FIGURE 6-11   (Part 2)

publications. Figure 6-12 shows the LLD factors for a number of General Electric lamps. Note that two factors are given in addition to the wattage. The Mean Lumen Factor is used when calculations involve average illumination between relampings (based upon lamp light output at 40 to 50 per cent of rated life, depending upon the type of lamp). The LLD factor is used when calculations involve minimum illumination between relampings (based on lamp light output at approximately 70 per cent of rated life). Some manufacturers use a *maintenance factor* (MF) or *depreciation factor* (DF) instead of LLD. No matter what factor is used, the effect of the lamp depreciation factor is to reduce the CU, thus increasing the number of luminaires for a given area and desired footcandles.

7. Look up the LDD. The LDD is sometimes specified in manufacturer's data for a given set of conditions and a given time period. In other cases the manufacturer will specify an IES maintenance category. Figure 6-13 shows the six IES maintenance categories. The LLD is specified for a given time period and a given set of dirt conditions.

For example, a category I lamp used for 48 months in a very clean area has an LDD factor of 90 per cent (Fig. 6-13). If the same lamp is used in a very dirty area for the same time, the LDD is about 53 per cent.

The extremes of very clean and very dirty are specified in other IES tables. The following rules of thumb can be applied for most practical work. *Very clean* is typical of high-grade offices, not near production, such as laboratories, clean rooms, and so on. *Clean* is found in offices of older buildings, or those near production, light assembly areas, and inspection areas. *Medium* is typical of mill offices, paper processing, and light machining. *Dirty* is found in heat treating, high-speed printing, and rubber processing. *Very dirty* is similar to dirty, but luminaires are within the immediate area of contamination (a constant accumulation of adhesive dirt). As an alternative, *very clean* is a condition of *no dirt whatsoever* (very unusual), *clean* is the cleanest conditions imaginable, *medium* is clean (but not the cleanest), *dirty* is dirty (but not the dirtiest), and *very dirty* is the dirtiest conditions imaginable.

8. Calculate the number of luminaires required using the infor-

GE Lamp Type	Watts	Mean Lumen Factor*(%)	LLD** (%)
Incandescent IOOA	100	93	90
IOOA	150	93	90
150PAR/SP and FL	150	84	78
150R/SP and FL	150	89	85
300M/IF	300	91	87
300R/SP and FL	300	94	92
500/IF	500	91	88
1000/IF	1000	92	89
1500/IF	1500	84	78
Fluorescent F40CW Mainlighter®	40	87	83
F40WW Mainlighter®	40	87	83
F40CWX Mainlighter®	40	83	73
F40WWX Mainlighter®	40	83	73
F40CW/S Staybright®	40	90	86
F40WW/S Staybright®	40	90	86
F96T12/CW	50	91	87
F96T12/WW	50	91	87
F96T12/CWX	50	87	78
F48T12/CW/HO	60	87	79
F96T12/CW/HO	110	87	77
F48PG17/CW	110	77	67
F96PG17/CW	215	75	65
F48T12/CW/1500	110	77	67
F96T12/CW/1500	215	80	71
High Intensity Discharge			
Mercury† H100DX38-4 (H38-4JA/DX)	100	76	69
H175DX39-22 (H39-22KC/DX)	175	81	76
H175WDX39-22 (H39-22KC/WDX)	175	77	71
H175RDXFL39-22 (H39-22BP/DX)	175	80	63
H250DX37-5 (H37-5KC/DX)	250	81	75
H250WDX37-5 (H37-5KC/WDX)	250	72	61
H400A33-1 (H33-1CD)	400	87	80
H400DX33-1 (H33-1GL/DX)	400	78	71
H400WDX33-1 (H33-1GL/WDX)	400	74	65
H1000A36-15 (H36-15GV)	1000	78	70
H1000DX36-15 (H36-15GW/DX)	1000	63	52
H1000WDX36-15 (H36-15GW/WDX)	1000	60	–
Multi-Vapor® MV400/BU/I and BD/I	400	78	70
MV1000BU/I	1000	80	75
Lucalox® LU250/BU and BD	250	91	88
LU400/BU and BD	400	90	84
LU1000/BU and BD	1000	92	90

* Use the Mean Lumen Factor when calculations involve average illumination between relampings (based upon lamp light output at 40%–50% of rated life depending upon lamp type).

** Use the LLD Factor when calculations involve *minimum* illumination between relampings (based upon lamp light output at approximately 70% of rated life).

† Mean Lumen Factors for mercury lamps are based upon 24,000 hours life.

Note: Factors shown for fluorescent lamps are based on 3 hours/start. Factors for Multi-Vapor Lucalox lamps are based on 10 hour/start.

FIGURE 6-12   Lamp lumen depreciation factors for General Electric lamps (*Courtesy* General Electric Company).

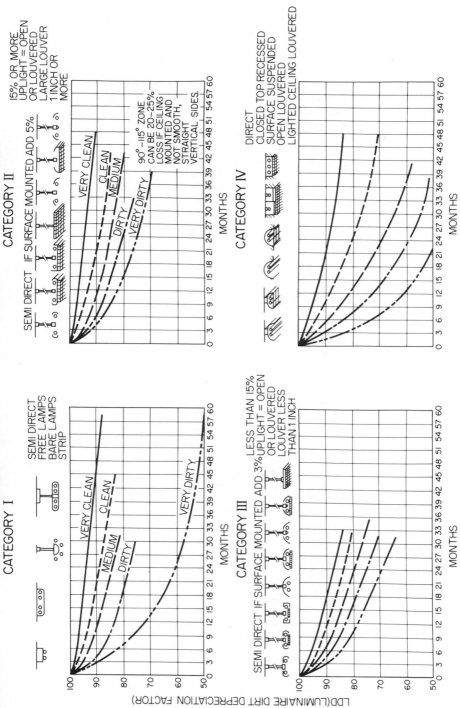

FIGURE 6-13 Luminaire dirt depreciation factors for six luminaire categories (I to VI) and for five degrees of dirtiness, per IES standards. (Part 1)

179

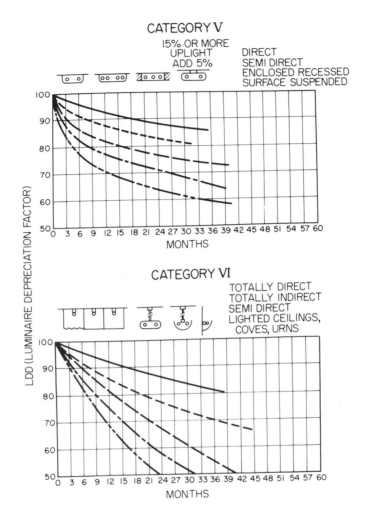

FIGURE 6-13 Luminaire dirt depreciation factors for six luminaire categories (I to VI) and for five degrees of dirtiness, per IES standards. (Part 2)

mation of Section 6-3.1. All the factors except CU can be calculated easily or looked up in the appropriate tables (manufacturer's data, IES standard data, etc.). Examples of how the number of luminaires can be calculated are given in Sections 6-3.4 and 6-3.5. Section 6-3.3 describes the procedure for calculating CU using the zonal cavity method.

6-3.3 Calculating CU

CU calculations involve three basis steps:

1. Determining the cavity *ratios*.
2. Determining the *effective* reflectances.
3. Looking up a CU from tables (IES or manufacturer's) using the ratios and reflectances.

The zonal cavity method derives its name from the technique of dividing a room into zones or cavities as part of the CU calculation procedure.

A room with suspended luminaires and a work place *not* on the floor has three cavities: ceiling cavity, room cavity, and floor cavity. Using surface-mounted or recessed luminaires eliminates the ceiling cavity. If the work plane is on the floor, there is no floor cavity. However, there is always a room cavity.

The cavities are shown on the work sheet of Fig. 6-14. This work sheet, developed by General Electric, reduces the CU calculation to a simple matter of filling in blanks. Note that h_{CC}, h_{RC}, and h_{FC} denote the height of the ceiling cavity, room cavity, and floor cavity (in feet), respectively. The symbol ρ means reflectance of the surface indicated (in per cent).

6-3.4 Example of zonal cavity method

The following paragraphs describe the step-by-step procedures for finding the number of luminaires (and luminaire spacing) for a given set of conditions using the zonal cavity method.

Determine the number of luminaires, and the spacing of luminaires, to provide an average maintained illumination of 150 fc in an industrial assembly area. The area dimensions are 50 X 50 X 25 (length, width, height) ft. The lamp to be used is a General Electric H400DX33-1 400-W deluxe white mercury lamp. The manufacturer describes the lamp as having 22,500 lm. The luminaire is described as an IES 24, one lamp per luminaire, with an LLD at relamping of 0.67. The dirt conditions are medium, with the luminaires cleaned every 2 years. The luminaires are mounted 21.5 ft from the floor. The work plane is 1.5 ft high. The maintained reflectances for the ceiling, walls, and floor are 80, 50, and 10 per cent, respectively.

COEFFICIENT OF UTILIZATION WORK SHEET

Job Identification: _Example - Industrial Assembly Area_

Name: _____ — _____ Date: _____ — _____

Average maintained footcandles: ___150___

        Luminaire Data: ___# 24___   Lamp Data: ___H 400 DX 33-1___

        Mfr: ___I E S___   Type: ___Mercury___

        Catalog No: ___—___   Lumens: ___22500___

        Lamps/Luminaire: ___1___   LLD: ___0.67___

        LDD: ___.80___   Lumens/Luminaire: ___22500___

Step 1. Fill in sketch

$$\underline{\quad 3.5 \quad} = h_{CC}$$

$$\underline{\quad 20 \quad} = h_{RC}$$

---Work plane---

$$\underline{\quad 1.5 \quad} = h_{FC}$$

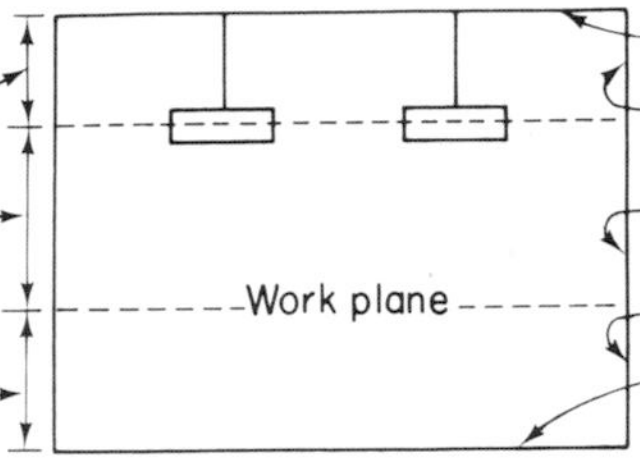

$\rho_C = 80$

$\rho_W = 50$

$\rho_W = 50$

$\rho_W = 50$

$\rho_F = 10$

Length: ___50___   Width: ___50___   Height: ___25___

Step 2. Determine cavity ratios from Fig. 6-15 or from formula

$$CR = \frac{5 \times h_C(L + W)}{L \times W}$$

Ceiling Cavity Ratio CCR = ___0.7___

Room Cavity Ratio RCR = ___4.0___

Floor Cavity Ratio FCR = ___0.3___

Step 3. Obtain Effective Ceiling Cavity Reflectance (ρ_{CC}) from Fig. 6-16 ρ_{CC} = ___70___

Obtain Effective Floor Cavity Reflectance (ρ_{FC}) from Fig. 6-16 ρ_{FC} = ___10___

Obtain Coefficient of Utilization (CU) from luminaire CU table CU = ___0.61___

If ρ_{FC} is other than 20% adjust CU by factor from Fig. 6-16

$$0.61 / 1.04 = 0.585$$

Adjusted CU = ___0.59___

FIGURE 6-14 Coefficients of utilization work sheet, long form (*Courtesy General Electric Company*).

1. Fill in all blanks of the work sheet (Fig. 6-14) possible, with the data of the stated problem. These include the following: The average maintained footcandles is specified as 150. The luminaire data are given as an IES 24. Figure 6-11 shows that a No. 24 has one lamp per luminaire. The lamp data are a GE H400DX33-1 mercury lamp with 22,500 lm and an LLD of

0.67.   Since there is one lamp per luminaire, each luminaire (lumens/luminaire) has 22,500 lm.   The length (50), width (50), height (50), and maintained reflectances $\rho C(80)$, $\rho W(50)$, and $\rho F(10)$ are all given.

2. Determine the height of all three cavities (h_{CC}, h_{RC}, and h_{FC}) using the supplied data, and fill in the appropriate blanks. The luminaires are mounted 21.5 ft from the floor, and the room is 25 ft high.   Thus the ceiling cavity, h_{CC} (or distance from luminaire to ceiling), is 3.5 ft (25 – 21.5 = 3.5).   The floor cavity, h_{FC} (or distance from work plane to floor), is 1.5 ft. The room cavity, h_{RC} (or distance from work plane to luminaire), is 20 ft (21.5 – 1.5 = 20).

3. Determine the cavity ratios using the room and cavity dimensions and fill in the appropriate blanks.   This can be done using the work sheet equation of Fig. 6-14 or the standard IES cavity ratio table (Fig. 6-15).   Let us try both methods.

Using the equation, the ceiling cavity dimension (h_C) is 3.5; thus the cavity ratio is

$$\frac{5 \times 3.5 \times (50 + 50)}{50 \times 50} = 0.7$$

Using the standard table (Fig. 6-15), find the width and length under the "Room Dimensions" column, then move across this row to the cavity depth of 3.5, to find a ratio of 0.7.   The floor cavity ratio FCR and room cavity ratio RCR can be found in the same row, under depths of 1.5 and 20, as 0.3 and 4.0, respectively.

4. Determine the effective ceiling and floor cavity reflectances using the data of Fig. 6-16, and fill in the appropriate blanks. For the ceiling find the 80 per cent ceiling reflectance column, then find the 50 per cent wall reflectance subcolumn, then move down this column to the ceiling cavity ratio of 0.7, to find a 70 per cent effective ceiling cavity reflectance. For the floor, find the 10 per cent floor reflectance column, then find the 50 per cent wall reflectance subcolumn, then move down this column to the floor cavity ratio of 0.3, to find a 10 per cent effective floor cavity reflectance.

5. Determine the CU using the data of Fig. 6-11 or similar manufacturer's data. Find the data for IES 24. Find the 70 per cent

Room dimensions		Cavity depth																		
Width	Length	1.0	1.5	2.0	2.5	3.0	3.5	4.0	5.0	6.0	7.0	8	9	10	11	12	14	16	20	25	30
8	8	1.2	1.9	2.5	3.1	3.7	4.4	5.0	6.2	7.5	8.8	10.0	11.2	12.5	—	—	—	—	—	—	—
	10	1.1	1.7	2.3	2.8	3.4	3.9	4.5	5.6	6.8	7.9	9.0	10.1	11.3	12.4	—	—	—	—	—	—
	14	1.0	1.5	2.0	2.5	2.9	3.4	3.9	4.9	5.9	6.9	7.9	8.8	9.8	10.8	11.8	—	—	—	—	—
	20	0.9	1.3	1.8	2.2	2.6	3.1	3.5	4.4	5.3	6.1	7.0	7.9	8.8	9.6	10.5	12.3	—	—	—	—
	30	0.8	1.2	1.6	2.0	2.4	2.8	3.2	4.0	4.8	5.5	6.3	7.1	7.9	8.7	9.5	11.1	—	—	—	—
	40	0.8	1.1	1.5	1.9	2.3	2.6	3.0	3.8	4.5	5.3	6.0	6.8	7.5	8.3	9.0	10.5	12.0	—	—	—
10	10	1.0	1.5	2.0	2.5	3.0	3.5	4.0	5.0	6.0	7.0	8.0	9.0	10.0	11.0	12.0	—	—	—	—	—
	14	0.9	1.3	1.7	2.1	2.6	3.0	3.4	4.3	5.1	6.0	6.9	7.7	8.6	9.4	10.3	12.0	—	—	—	—
	20	0.8	1.1	1.5	1.9	2.3	2.6	3.0	3.8	4.5	5.3	6.0	6.8	7.5	8.3	9.0	10.5	12.0	—	—	—
	30	0.7	1.0	1.3	1.7	2.0	2.3	2.7	3.3	4.0	4.7	5.3	6.0	6.7	7.3	8.0	9.3	10.7	—	—	—
	40	0.6	0.9	1.3	1.6	1.9	2.2	2.5	3.1	3.8	4.4	5.0	5.6	6.3	6.9	7.5	8.8	10.0	12.5	—	—
	60	0.6	0.9	1.2	1.5	1.8	2.0	2.3	2.9	3.5	4.1	4.7	5.3	5.8	6.4	7.0	8.2	9.3	11.7	—	—
12	12	0.8	1.3	1.7	2.1	2.5	2.9	3.3	4.2	5.0	5.8	6.7	7.5	8.3	9.2	10.0	11.7	—	—	—	—
	16	0.7	1.1	1.5	1.8	2.2	2.6	2.9	3.6	4.4	5.1	5.8	6.6	7.3	8.0	8.7	10.2	11.7	—	—	—
	24	0.6	0.9	1.3	1.6	1.9	2.2	2.5	3.1	3.8	4.4	5.0	5.6	6.3	6.9	7.5	8.8	10.0	12.5	—	—
	36	0.6	0.8	1.1	1.4	1.7	1.9	2.2	2.8	3.3	3.9	4.4	5.0	5.6	6.1	6.7	7.8	8.9	11.1	—	—
	50	0.5	0.8	1.0	1.3	1.6	1.8	2.1	2.6	3.1	3.6	4.1	4.7	5.2	5.7	6.2	7.2	8.3	10.3	—	—
	70	0.5	0.7	1.0	1.2	1.5	1.7	2.0	2.4	2.9	3.4	3.9	4.4	4.9	5.4	5.9	6.8	7.8	9.8	12.2	—
14	14	0.7	1.1	1.4	1.8	2.1	2.5	2.9	3.6	4.3	5.0	5.7	6.4	7.1	7.9	8.6	10.0	11.4	—	—	—
	20	0.6	0.9	1.2	1.5	1.8	2.1	2.4	3.0	3.6	4.3	4.9	5.5	6.1	6.7	7.3	8.5	9.7	12.1	—	—
	30	0.5	0.8	1.0	1.3	1.6	1.8	2.1	2.6	3.1	3.7	4.2	4.7	5.2	5.8	6.3	7.3	8.4	10.5	—	—
	42	0.5	0.7	1.0	1.2	1.4	1.7	1.9	2.4	2.9	3.3	3.8	4.3	4.8	5.2	5.7	6.7	7.6	9.5	11.9	—
	60	0.4	0.7	0.9	1.1	1.3	1.5	1.8	2.2	2.6	3.1	3.5	4.0	4.4	4.8	5.3	6.2	7.0	8.8	11.0	—
	90	0.4	0.6	0.8	1.0	1.2	1.4	1.7	2.1	2.5	2.9	3.3	3.7	4.1	4.5	5.0	5.8	6.6	8.3	10.3	12.4
17	17	0.6	0.9	1.2	1.5	1.8	2.1	2.4	2.9	3.5	4.1	4.7	5.3	5.9	6.5	7.1	8.2	9.4	11.8	—	—
	25	0.5	0.7	1.0	1.2	1.5	1.7	2.0	2.5	3.0	3.5	4.0	4.4	4.9	5.4	5.9	6.9	7.9	9.9	12.4	—
	35	0.4	0.7	0.9	1.1	1.3	1.5	1.7	2.2	2.6	3.1	3.5	3.9	4.4	4.8	5.2	6.1	7.0	8.7	10.9	—
	50	0.4	0.6	0.8	1.0	1.2	1.4	1.6	2.0	2.4	2.8	3.2	3.5	3.9	4.3	4.7	5.5	6.3	7.9	9.9	11.8
	80	0.4	0.5	0.7	0.9	1.1	1.2	1.4	1.8	2.1	2.5	2.9	3.2	3.6	3.9	4.3	5.0	5.7	7.1	8.9	10.7
	120	0.3	0.5	0.7	0.8	1.0	1.2	1.3	1.7	2.0	2.4	2.7	3.0	3.4	3.7	4.0	4.7	5.4	6.7	8.4	10.1
20	20	0.5	0.8	1.0	1.3	1.5	1.8	2.0	2.5	3.0	3.5	4.0	4.5	5.0	5.5	6.0	7.0	8.0	10.0	12.5	—
	30	0.4	0.6	0.8	1.0	1.3	1.5	1.7	2.1	2.5	2.9	3.3	3.8	4.2	4.6	5.0	5.8	6.7	8.3	10.4	12.5
	45	0.4	0.5	0.7	0.9	1.1	1.3	1.4	1.8	2.2	2.5	2.9	3.3	3.6	4.0	4.3	5.1	5.8	7.2	9.0	10.8
	60	0.3	0.5	0.7	0.8	1.0	1.2	1.3	1.7	2.0	2.3	2.7	3.0	3.3	3.7	4.0	4.7	5.3	6.7	8.3	10.0
	90	0.3	0.5	0.6	0.8	0.9	1.1	1.2	1.5	1.8	2.1	2.4	2.8	3.1	3.4	3.7	4.3	4.9	6.1	7.6	9.2
	150	0.3	0.4	0.6	0.7	0.9	1.0	1.1	1.4	1.7	2.0	2.3	2.6	2.8	3.1	3.4	4.0	4.5	5.7	7.1	8.5
24	24	0.4	0.6	0.8	1.0	1.3	1.5	1.7	2.1	2.5	2.9	3.3	3.8	4.2	4.6	5.0	5.8	6.7	8.3	10.4	12.5
	32	0.4	0.5	0.7	0.9	1.1	1.3	1.5	1.8	2.2	2.6	2.9	3.3	3.6	4.0	4.4	5.1	5.8	7.3	9.1	10.9
	50	0.3	0.5	0.6	0.8	0.9	1.1	1.2	1.5	1.9	2.2	2.5	2.8	3.1	3.4	3.7	4.3	4.9	6.2	7.7	9.2
	70	0.3	0.4	0.6	0.7	0.8	1.0	1.1	1.4	1.7	2.0	2.2	2.5	2.8	3.1	3.4	3.9	4.5	5.6	7.0	8.4
	100	0.3	0.4	0.5	0.6	0.8	0.9	1.0	1.3	1.6	1.8	2.1	2.3	2.6	2.8	3.1	3.6	4.1	5.2	6.5	7.8
	160	0.2	0.4	0.5	0.6	0.7	0.8	1.0	1.2	1.4	1.7	1.9	2.2	2.4	2.6	2.9	3.4	3.8	4.8	6.0	7.2

FIGURE 6-15   Cavity ratios, per IES standards. (Part 1)

FIGURE 6-15 Cavity ratios, per IES standards. (Part 2)

Room dimensions		Cavity depth																			
Width	Length	1.0	1.5	2.0	2.5	3.0	3.5	4.0	5.0	6.0	7.0	8	9	10	11	12	14	16	20	25	30
30	30	0.3	0.5	0.7	0.8	1.0	1.2	1.3	1.7	2.0	2.3	2.7	3.0	3.3	3.7	4.0	4.7	5.4	6.7	8.4	10.0
	45	0.3	0.4	0.6	0.7	0.8	1.0	1.1	1.4	1.7	1.9	2.2	2.5	2.7	3.0	3.3	3.8	4.4	5.5	6.9	8.2
	60	0.3	0.4	0.5	0.6	0.7	0.9	1.0	1.2	1.5	1.7	2.0	2.2	2.5	2.7	3.0	3.5	4.0	5.0	6.2	7.4
	90	0.2	0.3	0.4	0.6	0.6	0.8	0.9	1.1	1.3	1.6	1.8	2.0	2.2	2.5	2.7	3.1	3.6	4.5	5.6	6.7
	150	0.2	0.3	0.4	0.5	0.6	0.7	0.8	1.0	1.2	1.4	1.6	1.8	2.0	2.2	2.4	2.8	3.2	4.0	5.0	5.9
	200	0.2	0.3	0.4	0.5	0.6	0.7	0.8	1.0	1.1	1.3	1.5	1.7	1.9	2.0	2.2	2.6	3.0	3.7	4.7	5.6
36	36	0.3	0.4	0.6	0.7	0.8	1.0	1.1	1.4	1.7	1.9	2.2	2.5	2.8	3.0	3.3	3.8	4.1	5.5	6.9	8.3
	50	0.2	0.4	0.5	0.6	0.7	0.8	1.0	1.2	1.4	1.7	1.9	2.1	2.5	2.6	2.9	3.3	3.8	4.8	5.9	7.2
	75	0.2	0.3	0.4	0.5	0.6	0.7	0.8	1.0	1.2	1.4	1.6	1.8	2.0	2.3	2.5	2.9	3.3	4.1	5.1	6.1
	100	0.2	0.2	0.4	0.5	0.6	0.7	0.8	0.9	1.1	1.3	1.5	1.7	2.0	2.1	2.1	2.6	3.0	3.8	4.7	5.7
	150	0.2	0.3	0.3	0.4	0.5	0.6	0.7	0.9	1.0	1.2	1.4	1.6	1.7	1.9	2.1	2.4	2.8	3.5	4.3	5.2
	200	0.2	0.2	0.3	0.4	0.5	0.6	0.7	0.8	1.0	1.1	1.3	1.5	1.6	1.8	2.0	2.3	2.6	3.3	4.1	4.9
42	42	0.2	0.4	0.5	0.6	0.7	0.8	1.0	1.2	1.4	1.6	1.9	2.1	2.4	2.6	2.8	3.3	3.8	4.7	5.9	7.1
	60	0.2	0.3	0.4	0.5	0.6	0.7	0.8	1.0	1.2	1.4	1.6	1.8	2.0	2.2	2.4	2.8	3.2	4.0	5.0	6.0
	90	0.2	0.2	0.3	0.4	0.5	0.6	0.7	0.9	1.0	1.2	1.4	1.6	1.7	1.9	2.1	2.4	2.8	3.5	4.4	5.2
	140	0.1	0.2	0.3	0.3	0.5	0.5	0.6	0.8	0.9	1.1	1.2	1.4	1.5	1.7	1.9	2.2	2.5	3.1	3.9	4.6
	200	0.1	0.2	0.3	0.3	0.4	0.5	0.6	0.7	0.9	1.0	1.1	1.3	1.4	1.6	1.7	2.0	2.3	2.9	3.6	4.3
	300	0.1	0.2	0.2	0.3	0.4	0.4	0.5	0.7	0.8	0.9	1.1	1.3	1.4	1.5	1.7	1.9	2.2	2.7	3.5	4.2
50	50	0.2	0.3	0.4	0.5	0.6	0.7	0.8	1.0	1.2	1.4	1.6	1.8	2.0	2.2	2.4	2.8	3.2	4.0	5.0	6.0
	70	0.2	0.3	0.3	0.4	0.5	0.6	0.7	0.9	1.0	1.2	1.4	1.5	1.7	1.9	2.0	2.4	2.7	3.4	4.3	5.1
	100	0.1	0.2	0.3	0.4	0.4	0.5	0.6	0.7	0.9	1.0	1.2	1.3	1.5	1.6	1.8	2.1	2.4	3.0	3.7	4.5
	150	0.1	0.2	0.3	0.3	0.4	0.5	0.6	0.7	0.8	0.9	1.1	1.2	1.3	1.5	1.6	1.9	2.1	2.7	3.3	4.0
	300	0.1	0.2	0.2	0.3	0.4	0.4	0.5	0.6	0.7	0.8	0.9	1.0	1.1	1.3	1.4	1.6	1.9	2.3	2.9	3.5
60	60	0.2	0.2	0.3	0.4	0.5	0.6	0.7	0.8	1.0	1.2	1.3	1.5	1.7	1.8	2.0	2.3	2.7	3.3	4.2	5.0
	100	0.1	0.2	0.3	0.3	0.4	0.5	0.5	0.7	0.8	0.9	1.1	1.2	1.3	1.5	1.6	1.9	2.1	2.7	3.3	4.0
	150	0.1	0.2	0.2	0.3	0.3	0.4	0.5	0.6	0.7	0.8	0.9	1.0	1.2	1.3	1.4	1.6	1.9	2.3	2.9	3.5
	300	0.1	0.1	0.2	0.2	0.3	0.3	0.4	0.5	0.6	0.7	0.8	0.9	1.0	1.1	1.2	1.4	1.6	2.0	2.5	3.0
75	75	0.1	0.2	0.3	0.3	0.4	0.5	0.5	0.7	0.8	0.9	1.1	1.2	1.3	1.5	1.6	1.9	2.1	2.7	3.3	4.0
	120	0.1	0.2	0.2	0.3	0.3	0.4	0.4	0.5	0.7	0.8	0.9	1.0	1.1	1.2	1.3	1.5	1.7	2.2	2.7	3.3
	200	0.1	0.1	0.2	0.2	0.3	0.3	0.4	0.5	0.6	0.6	0.7	0.8	0.9	1.0	1.1	1.3	1.5	1.8	2.3	2.7
	300	0.1	0.1	0.2	0.2	0.3	0.3	0.3	0.4	0.5	0.6	0.7	0.8	0.8	0.9	1.0	1.2	1.3	1.7	2.1	2.5
100	100	0.1	0.2	0.2	0.2	0.3	0.3	0.4	0.5	0.6	0.7	0.8	0.9	1.0	1.1	1.2	1.4	1.6	2.0	2.5	3.0
	200	0.1	0.1	0.2	0.2	0.2	0.3	0.3	0.4	0.5	0.5	0.6	0.7	0.8	0.8	0.9	1.1	1.2	1.5	1.9	2.2
	300	0.1	0.1	0.1	0.2	0.2	0.2	0.3	0.3	0.4	0.5	0.5	0.6	0.7	0.7	0.8	0.9	1.1	1.3	1.7	2.0
150	150	0.1	0.1	0.1	0.2	0.2	0.2	0.3	0.3	0.4	0.5	0.5	0.6	0.7	0.7	0.8	0.9	1.1	1.3	1.7	2.0
	300	0.1	0.1	0.1	0.1	0.2	0.2	0.2	0.3	0.3	0.4	0.4	0.5	0.5	0.6	0.6	0.7	0.8	1.0	1.2	1.5
200	200	0.1	0.1	0.1	0.1	0.2	0.2	0.2	0.3	0.3	0.4	0.4	0.5	0.5	0.6	0.6	0.7	0.8	1.0	1.2	1.5
	300	0.1	0.1	0.1	0.1	0.1	0.1	0.2	0.2	0.3	0.3	0.3	0.4	0.4	0.5	0.5	0.6	0.7	0.8	1.0	1.2
300	300	—	0.1	0.1	0.1	0.1	0.1	0.1	0.2	0.2	0.2	0.3	0.3	0.3	0.4	0.4	0.5	0.5	0.7	0.8	1.0
	500	—	—	0.1	0.1	0.1	0.1	0.1	0.1	0.2	0.2	0.2	0.2	0.3	0.3	0.3	0.4	0.4	0.5	0.7	0.8
500	500	—	—	—	0.1	0.1	0.1	0.1	0.1	0.1	0.1	0.2	0.2	0.2	0.2	0.2	0.3	0.3	0.4	0.5	0.6

Ceiling or floor cavity ratio	90	90	90	90	(80)	80	(50)	30	70	70	70	50	50	50	30	30	30	30	(10)	(10)	(10)
Per cent wall reflectance →	90	70	50	30	80	70	(50)	30	70	50	30	70	50	30	65	50	30	10	(50)	30	10
0	90	90	90	90	80	80	80	80	70	70	70	50	50	50	30	30	30	30	10	10	10
0.1	90	89	88	87	80	79	78	78	69	69	68	50	49	48	30	30	30	29	10	10	10
0.2	89	88	86	85	79	79	78	76	68	67	66	49	49	47	30	29	29	28	10	10	10
0.3	89	87	85	83	79	78	77	74	67	66	64	49	48	47	30	29	29	28	(10)	10	9
0.4	88	86	83	81	78	78	75	72	67	65	63	48	47	46	30	29	27	27	11	10	9
0.5	88	85	81	78	77	77	74	70	66	64	61	48	46	45	29	28	27	26	11	10	9
0.6	88	84	80	76	77	75	71	68	65	62	59	47	45	43	29	28	26	25	11	10	9
0.7	88	83	78	74	76	74	(70)	66	65	61	58	47	44	42	29	28	26	24	12	10	8
0.8	87	82	77	73	75	73	69	65	64	60	56	47	43	41	29	27	25	23	12	9	8
0.9	87	81	76	71	75	72	68	63	63	59	55	46	43	40	29	27	25	22	12	9	8
1.0	86	80	74	69	74	71	66	61	63	58	53	46	42	39	29	27	24	22	12	9	8
1.1	86	79	73	67	74	71	65	60	62	57	52	46	41	38	29	26	24	21	12	9	8
1.2	86	78	72	65	73	70	64	58	61	56	50	45	41	37	29	26	23	20	12	9	7
1.3	85	78	71	64	73	68	63	57	60	55	49	45	40	37	28	26	23	20	12	9	7
1.4	85	77	69	62	72	68	62	55	60	54	48	45	40	35	28	26	22	19	12	9	7
1.5	85	76	68	61	72	68	61	54	59	53	47	44	39	34	28	25	22	18	12	9	7
1.6	85	75	66	59	71	67	60	53	59	52	45	46	41	38	29	25	21	18	12	9	7
1.7	84	74	65	58	71	66	59	52	58	51	44	45	41	37	29	25	21	17	12	9	7
1.8	84	73	64	56	70	65	58	50	57	50	43	43	40	32	28	25	21	17	12	9	6
1.9	84	73	63	55	70	65	57	49	57	49	42	43	40	31	28	25	21	16	12	9	6
2.0	83	72	62	53	69	64	56	48	56	48	41	43	39	30	28	24	20	16	12	9	6
2.1	83	71	61	52	69	63	55	47	56	47	40	43	36	29	28	24	20	16	12	9	6
2.2	83	70	60	51	68	63	54	45	55	46	39	42	36	29	28	24	19	15	13	9	6
2.3	83	69	59	50	68	62	53	44	53	46	38	42	35	28	28	24	19	15	13	9	6
2.4	82	68	58	48	67	61	52	43	54	45	37	42	35	27	28	24	19	14	13	9	6
2.5	82	68	57	47	67	61	51	42	53	44	36	41	34	27	27	23	18	14	13	9	6

FIGURE 6-16   Per cent effective ceiling or floor cavity reflectance for various reflectance combinations, per IES standards. (Part 1)

186

Per cent ceiling or floor reflectance	90				80				70			50			30				10		
Per cent wall reflectance	90	70	50	30	80	70	50	30	70	50	30	70	50	30	65	50	30	10	50	30	10
Ceiling or floor cavity ratio																					
2.6	82	67	56	46	66	60	50	41	53	43	35	41	34	26	27	23	18	13	13	9	5
2.7	82	66	55	45	66	60	49	40	52	43	34	41	33	26	27	23	18	13	13	9	5
2.8	81	66	54	44	66	59	48	39	52	42	33	41	33	25	27	23	18	13	13	9	5
2.9	81	65	53	43	65	58	48	38	51	41	33	40	33	25	27	23	17	12	13	9	5
3.0	81	64	52	42	65	58	47	38	51	40	32	40	32	24	27	22	17	12	13	8	5
3.1	80	64	51	41	64	57	46	37	50	40	31	40	32	24	27	22	17	12	13	8	5
3.2	80	63	50	40	64	57	45	36	50	39	30	40	31	23	27	22	16	11	13	8	5
3.3	80	62	49	39	64	56	44	35	49	39	30	39	31	23	27	22	16	11	13	8	5
3.4	80	62	48	38	63	56	44	34	49	38	29	39	31	22	27	22	16	11	13	8	5
3.5	79	61	48	37	63	55	43	33	48	38	29	39	30	22	26	22	16	11	13	8	5
3.6	79	60	47	36	62	54	42	33	48	37	28	39	30	21	26	21	15	10	13	8	5
3.7	79	60	46	35	62	54	42	32	48	37	27	39	30	21	26	21	15	10	13	8	4
3.8	79	59	45	35	62	53	41	31	47	36	27	38	29	21	26	21	15	10	13	8	4
3.9	78	59	45	34	61	53	40	30	47	36	26	38	29	20	26	21	15	10	13	8	4
4.0	78	58	44	33	61	52	40	30	46	35	26	38	29	20	26	21	15	9	13	8	4
4.1	78	57	43	32	60	52	39	29	46	35	25	37	28	20	26	21	14	9	13	8	4
4.2	78	57	43	32	60	51	39	29	46	34	25	37	28	19	26	20	14	9	13	8	4
4.3	78	56	42	31	60	51	38	28	45	34	25	37	28	19	26	20	14	9	13	8	4
4.4	77	56	41	30	59	51	38	28	45	34	24	37	27	19	26	20	14	9	13	8	4
4.5	77	55	41	30	59	51	37	27	45	33	24	37	27	19	25	20	14	8	14	8	4
4.6	77	55	40	29	59	50	37	26	44	33	24	36	27	18	25	20	14	8	14	8	4
4.7	77	54	40	29	58	49	36	26	44	33	23	36	26	18	25	20	13	8	14	8	4
4.8	76	54	39	28	58	49	36	25	44	32	23	36	26	18	25	19	13	8	14	8	4
4.9	76	53	38	28	58	49	35	25	44	32	23	36	26	18	25	19	13	7	14	8	4
5.0	76	53	38	27	57	48	35	25	43	32	22	36	26	17	25	19	13	7	14	8	4

FIGURE 6-16   (Part 2)

effective ceiling cavity reflectance column, then find the 50 per cent wall reflectance subcolumn, then move down this column to the room cavity ratio of 4, to find a CU of 0.61.

Note that all the information in Fig. 6-11 is based on a CU for 20 per cent effective floor cavity reflectance. In our example floor cavity reflectance is 10 per cent, as determined in step 4. The CU must be corrected to show the difference in floor reflectance using the data of Fig. 6-17. Find the 70 per cent effective ceiling cavity reflectance column, then find the 50 per cent wall reflectance subcolumn, then move down this column to the room cavity ratio of 4, to find a correction factor of 1.04. Divide the CU of 0.61 by the correction factor of 1.04, to find an adjusted CU of 0.59.

6. The final factor to be determined is LDD. Figure 6-11 shows that IES 24 is LDD maintenance category III. Figure 6-13 shows that at 2 years (24 months), with medium dirt conditions, the LDD for category III is 80 per cent.
7. With all the factors on the work sheet, find the number of luminaires using the data of section 6-3.1.

$$\text{number of luminaires} = \frac{150 \times 2500}{0.59 \times 22500 \times 1 \times 0.80 \times 0.67} = 53$$

For 30 per cent effective floor cavity reflectance, multiply by appropriate factor below.
For 10 per cent effective floor cavity reflectance, divide by appropriate factor below.

Per cent effective ceiling cavity reflectance, ρ_{cc}	80			70			50			10		
Per cent wall reflectance, ρ_w	50	30	10	50	30	10	50	30	10	50	30	10
Room cavity ratio												
1	1.08	1.08	1.07	1.07	1.06	1.06	1.05	1.04	1.04	1.01	1.01	1.0
2	1.07	1.06	1.05	1.06	1.05	1.04	1.04	1.03	1.03	1.01	1.01	1.0
3	1.05	1.04	1.03	1.05	1.04	1.03	1.03	1.03	1.02	1.01	1.01	1.0
4	1.05	1.03	1.02	1.04	1.03	1.02	1.03	1.02	1.02	1.01	1.01	1.0
5	1.04	1.03	1.02	1.03	1.02	1.02	1.02	1.02	1.01	1.01	1.01	1.00
6	1.03	1.02	1.01	1.03	1.02	1.01	1.02	1.02	1.01	1.01	1.01	1.00
7	1.03	1.02	1.01	1.03	1.02	1.01	1.02	1.01	1.01	1.01	1.01	1.0
8	1.03	1.02	1.01	1.02	1.02	1.01	1.02	1.01	1.01	1.01	1.01	1.0
9	1.02	1.01	1.01	1.02	1.01	1.01	1.02	1.01	1.01	1.01	1.01	1.0
10	1.02	1.01	1.01	1.02	1.01	1.01	1.02	1.01	1.01	1.01	1.01	1.0

FIGURE 6-17   Factors for effective floor cavity reflectance other than 20 per cent, per IES standards.

6-3.4.1 **Luminaire spacing.** Once the number of luminaires has been established, the next step is to decide on a practical plan for spacing the luminaires so that the light will be uniform. Figure 6-11 shows that an IES 24 has a maximum spacing/mounting height above the work plane ratio of 1.4. If this ratio is exceeded, the lighting will be nonuniform. The maximum spacing is 1.4 X 20 (distance between work plane and luminaire), or 28, ft. Obviously, uniformity will be no problem in this 50 X 50 foot space.

Divide the number of luminaires into the area (2500/53) to find an area of 47 ft ^{2} per fixture. Six by 8 ft is thus a reasonable spacing or six rows of nine fixtures each, each spaced to fit. Note that this provides one extra luminaire.

The following notes, based on IES recommendations, may prove helpful in planning the spacing of luminaires. The illumination levels recommended by most manufacturers and the IES are *minimum foot-candles on* the task being performed at the work area. In most in-stances the task is likely to be located (or relocated) at any point in the room being lighted. Logically, the illumination at any point should not differ materially from the calculated average. Perfectly uniform illumination is simply not practical in all but a few cases. However, uniformity is considered acceptable if the maximum and minimum values in the room are not more than one-sixth above, and one-sixth below, the average.

For acceptable uniformity, luminaires should not be spaced too far apart or to far from the walls. Spacing limitations between direct, semidirect, and general-diffuse luminaires are related to the mounting height above the work plane. For semiindirect and indirect luminaires, the ceiling height above the work plane is the dimension of reference.

Recommended spacing/mounting or ceiling height above the work plane ratios are given in Fig. 6-11 directly below the light distribution curve for each luminaire type. These spacing ratios apply to the spacings shown in Fig. 6-18.

The commonly used practice of letting the distance from the lumi-naires to the wall equal one-half the distance between rows (see Fig. 6-19(a) results in inadequate illumination near the walls. Since desks and benches are frequently located along the walls, a distance of $2\frac{1}{2}$ ft from the wall to the center of the luminaire should be used to avoid excessive dropoff in illumination. This will locate the lumi-naires over the edge of desks facing the wall or over the center of desks

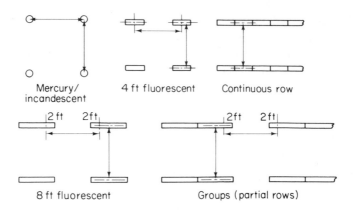

FIGURE 6-18   Spacing dimensions to be used in relation to spacing-to-mounting height ratios, per IES standards.

that are perpendicular to the wall (see Fig. 6-19b). To further improve illumination uniformity across the room, it is often desirable to use somewhat closer spacings between outer rows of luminaires than between center rows, taking care to be sure that no spacing exceeds the maximum permissible spacing.

To prevent excessive reduction in illumination at the ends of the room, the ends of fluorescent luminaire rows should preferably be 6 to 12 in. from the walls or in no case more than 2 ft from the walls. Even the 6- to 12-in. spacing leaves much to be desired from the standpoint of uniformity.   Where practical, the arrangement of Fig. 6-19c is much more satisfactory. With this arrangement the units at each end of the row are replaced by a continuous row parallel to, and $2\frac{1}{2}$ ft from, the end wall. In the example shown, five units were replaced by nine units, providing a potential increase in the illumination at the end of the room of 80 per cent over what it would be with the layout shown in Fig. 6-19b. This technique not only improves uniformity but also eliminates scallops of light on the end walls and provides a uniform wash of light on all four walls.

Another method of compensating for the normal reduction in illumination that may be expected at the ends of rows is to use a greater number of lamps in the end units. Still another technique is to provide additional units between the rows at each end. The units can be either parallel, or at right angles, to the rows.

Spacings closer than the maximum permissible are often highly desirable to reduce harsh shadows and ceiling reflections in the task.

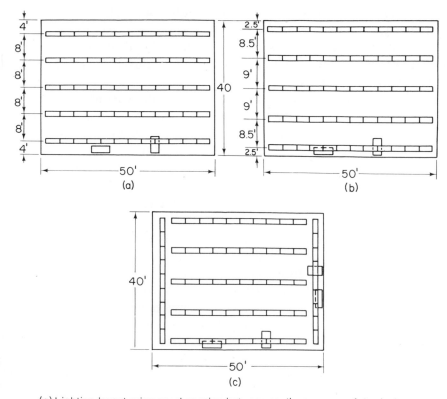

(a) Lighting layout using equal spacing between continuous rows of luminaires.
(b) Layout is changed to provide more illumination near side walls.
(c) By adding four more four-foot units on each end, layout (b) can be modified to provide 80 per cent more light near the end walls and prevent possible scallop effects.

FIGURE 6-19   Methods of luminaire spacing, per IES standards.

Close spacing also improves uniformity. This is particularly true for direct and semidirect equipment. Spacings that are substantially less than the maximum permissible spacing should be seriously considered (when cost permits).

6-3.5 Rules of thumb for estimates in lighting design

When all the information described in Section 6-3.4 is not available or when the figures do not work out to exact values, it is often necessary to make an "educated guess." In fact, it is often possible to produce a rough (but sufficiently accurate) estimate of lighting requirements if a few simple rules are followed. The following is a summary of these rules or guidelines.

When reflectance values and cavity ratios do not come out in even numbers that appear in tables, estimate the values. Often the intermediate values can be accurately determined by inspection, by examining the nearby values in the tables. Often what appears to be a large change in a particular factor will result in a very small difference in the number of luminaires. In those cases where accuracy is important, go through a mathematical interpolation.

When tables of CU are not available for the specific fixture picked from a catalog, use the standard IES tables of Fig. 6-11.

When no tables of any kind are available, remember the basic equation: lumens required = (footcandles X area)/CU. A very rough estimate can be found by assuming a CU. When all else fails, assume a CU of 0.5.   From a practical standpoint, most areas that require lighting will have CUs from about 0.4 to 0.6. If the space is unusually small, has low reflectance surfaces, or has better than average shielding on the luminaires (in effect, if the area is normally dark), use a lower CU value (0.4, or possibly less, for very dark areas that the customer wants to make light).   If the surfaces are light, the area is large with few partitions, and there are efficient lenses or diffusers on the luminaire, use a higher CU (0.6 or higher). The final estimate will be a guess, but it will be an "educated guess."

When the reflectances of the surfaces are unknown, use a light meter. The type and accuracy of the light meter is not too important as long as the readings are consistent. Measure the light falling on the unknown surface (ceiling, wall, or floor), then turn the meter around, move the meter 3 to 4 in. from the surface, and measure the light being reflected from the surface. The reflected light, divided by the incident light, times 100, is the *approximate* reflectance of the surface in per cent. The equation is: per cent of reflectance = (reflected light/incident light) X 100.

When lamp catalogs are not available and the lumen output of various lamps is needed, a rough estimate can be obtained by memorizing general lamp lumens-per-watt characteristics and multiplying by the wattage of the lamp to be used. Remember the following:

Incandescent   20 lm/W (50% less if PAR or R lamp)
Mercury   50 lm/W
Fluorescent   80 lm/W

Multivapor   85 lm/W
Lucalox   100 lm/W

6-3.5.1 Example of quick-estimate method. The following paragraphs describe the step-by-step procedures for finding the number
of luminaires for a given set of conditions using the quick-estimate
method. Solve the following problem two ways: first with the quick
estimate and second by working through the calculations using the
procedures of Section 6-3.4.

The lobby of an office building is to be used as a display area for
large industrial equipment. Find the number of luminaires required
to provide an average *initial* (not maintained) illumination level of
100 fc on the display equipment's top surfaces, which are 9 ft from
the floor. Use a recessed incandescent downlight (IES 15) equipped
with a 300-W R-40 lamp. The dimensions of the room are (L X W X H)
90 X 30 X 18 ft. Reflectances are (ceiling, walls, floor) 80, 30 and
30 per cent.

The surfaces are light, the room is fairly large, and the fixture
has no louvers or dark shielding, so efficiency is high, typically CU
0.6 or greater. Try a CU of 0.7. The lamp is incandescent; thus 20
lm/W, times 300 W, gives 6000 lm per lamp. However, light is trapped
inside the reflector and projector, so the value is reduced 50 per cent,
to 3000 lm per lamp. The area is 2700 ft^2 (90 X 30). Using the equation, lumens required = (fc X area)/CU, or (100 X 2700)/0.7 = 385,000,
3000 lm per lamp divided into 385,000 gives 128 fixtures.

Now calculate the solution according to the procedures of Section
6-3.4, using the work sheet of Fig. 6-20. The CU is found to be 0.67.
Substituting the work sheet values into the equation, the number of
luminaires is

$$\frac{100 \text{ X } 2700}{0.67 \text{ X } 3650 \text{ X } 1 \text{ X } 1 \text{ X } 1} = 110$$

Note that the problem relates to initial illumination rather than
maintained illumination. Thus LLD and LDD are equal to 1. Also,
the number of lamps per luminaires is 1. Thus the divisor of the equation can be found simply when the CU is multiplied by the lumens
per lamp. Of course, in this case knowing the actual lumens per lamp
would have made the estimated solution agree quite well with the
calculated solution.

COEFFICIENT OF UTILIZATION WORK SHEET

Job Identification: _Example – Lobby/Display Area_

Name: _____ — _____   Date: ___—___

Average initial footcandles: ___100___

Luminaire Data: # _15_   Lamp Data: _300R/FL_

Mfr: ___IES___   Type: _Incandescent_

Catalog No: ___—___   Lumens: ___3650___

Lamps/Luminaire: _1_   LLD: ___1.0___

LDD: ___1.0___   Lumens/Luminaire: _3650_

Step 1.   Fill in sketch

___0___ = h_{cc}

___9___ = h_{RC}

------Work plane------

___9___ = h_{FC}

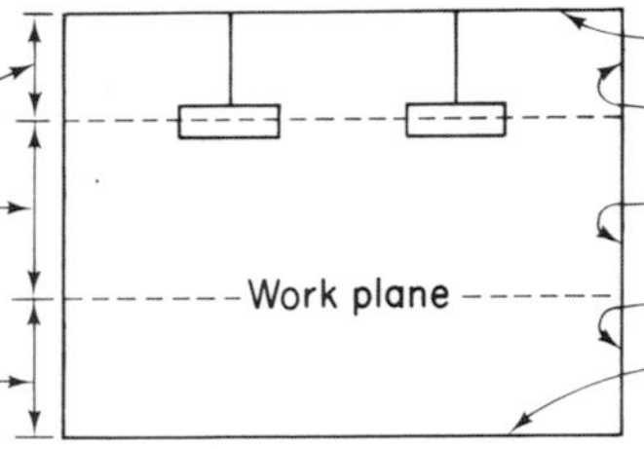

ρ_C = _80_
ρ_W = _30_
ρ_W = _30_
ρ_W = _30_
ρ_F = _30_

Length: _90_   Width: _30_   Height: _18_

Step 2.   Determine cavity ratios from Fig. 6–15 or from formula

$$CR = \frac{5 \times h_c(L + W)}{L \times W}$$

Ceiling Cavity Ratio CCR = ___0___
Room Cavity Ratio RCR = _2.0_
Floor Cavity Ratio FCR = _2.0_

Step 3.   Obtain Effective Ceiling Cavity Reflectance (ρ_{CC}) from Fig. 6–16 ρ_{CC} = _80_

Obtain Effective Floor Cavity Reflectance (ρ_{FC}) from Fig. 6–16 ρ_{FC} = _20_

Obtain Coefficient of Utilization (CU) from luminaire CU table CU = ___0.67___

If ρ_{FC} is other than 20% adjust CU by factor from Fig. 6–17

Adjusted CU = ___—___

FIGURE 6-20   Coefficient of utilization work sheet, short form (*Courtesy* **General Electric Company**).

6-4 LIGHTING CONTROL EQUIPMENT

The contactors (also called starters or controllers) for electric motors can sometimes be used to control lighting circuits. A description of such contactors is provided in Section 8-7.2.

Chapter 7

Electric Comfort
Heating Design

In this chapter we shall discuss the design of electric comfort heating systems from the standpoint of providing a given amount of heat for a given set of conditions. That is, we shall discuss calculations to determine how much heat is required for a given room or area, of given construction (building materials, construction methods, etc.), to maintain a given temperature.

This discussion is made using terms most familiar to those in the electrical industry. For example, British thermal units (Btu) are used in the heating field to express the quantity of heat. However, electrical heating contractors are more familiar with watts and kilowatts than with Btus. Therefore, the information in this chapter is expressed in watts wherever possible.

With the wattage established for a given heating requirement, the electrical wiring system (conductor size, voltage drop, transformer connections, etc.) for the heaters can be calculated using the information of Chapters 1 through 5.

The success of every electric comfort heating application boils down to four basic factors: adequate insulation, proper provisions for control of moisture, accurate heat-loss calculations, and proper installation of the electric heating system. All these areas are discussed in this chapter.

7-1 TYPES OF ELECTRIC COMFORT HEATING EQUIPMENT

The most common types of electric comfort heating equipment in use today are embedded radiant heating wire, wall heaters, baseboard heaters, utility heaters, and ceiling heaters. These heaters can be controlled by switches but generally control is by means of thermostats (or combination switch and thermostats). Heaters can be controlled by line voltage switches and thermostats or by controls that operate on low voltages (using relays and transformers as described in Chapter 5). Modern electric heating also includes humidity controls. The following paragraphs provide brief descriptions of these devices. Mechanical installation and wiring details are kept to a minimum except where necessary for planning and design of the system.

7-1.1 Embedded radiant heating wire

Each heating unit consists of a definite length of heating wire spliced at the factory to nonheating power leads. The heating wire and nonheating power leads *must not be cut, shortened, or altered in any manner.* This is most important since each heating unit is cut at the factory for a specified wattage at a rated voltage.

The nonheating power lead wires contain metal identification tags. These must remain intact after the installation has been completed.

Many different alloys are used in the production of heating wire. Because of manufacturing tolerances, the resistance of the alloy varies over a small range. Therefore, the length of each radiant heating unit is properly determined in the factory to provide the rated wattage at a specified voltage. The length listed for a given rating in the manufacturer's catalog is the actual length, and it should be used in determining spacing arrangement. (The procedures for finding the best spacing of radiant heating wires are discussed in later paragraphs of this section.)

Typically, radiant heating systems operate on 120, 208, or 240 V. As usual, 240 V requires the least current for a given wattage. Generally, the voltage rating of each unit is shown on the nonheating power lead tags and the spool containing the heating wire unit. For example, General Electric red colored nonheating power leads identify 240- and 208-V units; yellow leads identify 120-V wire.

Radiant heating can be installed in plastered ceilings, laminated gypsum wall board ceilings, and concrete floors. The procedures are different for each of the three installations. Figures 7-1, 7-2, and 7-3 illustrate the *basic requirements* for the three installations. Keep in mind when studying these illustrations that local inspection *must* be made before the concrete is poured or before the plaster is applied. Equally important, any serious miscalculation (improper spacing of wires, wires too close to fixtures, etc.) must be caught before the installation is complete.

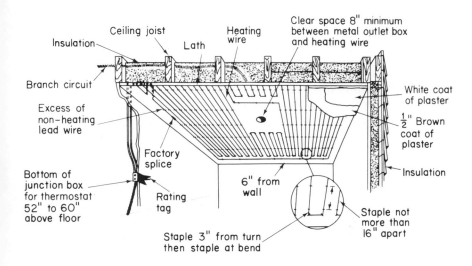

FIGURE 7-1   Basic spacing requirements for radiant heating wire in plaster ceilings (*Courtesy* General Electric Company).

Plaster ceiling installations. In addition to the requirements of Fig. 7-1, it is necessary to determine the required spacing between the heating wire runs. This can be done using the equation

$$\text{wire spacing} = \frac{12(L-1)(W-1) - 12A}{U}$$

where L is the length of the room (in ft); W is the width of the room (in ft); A is the total square feet of *unusable* ceiling area, such as that taken up by recessed lighting fixtures, closets, and kitchen cabinets; and U is the length of the heating wire unit (in ft).

Note: Never install heating wire in a plastered ceiling with a wire spacing less than $1\frac{1}{2}$ inches.

Example of plaster ceiling wire spacing

A room 11 X 12 ft requires 1.2 kW of embedded heating wire. (This value has been found using the data of Sections 7-5 and/or 7-6.) General Electric provides 1.2 kW at 240 V with a wire 444 ft long. (Select the proper wire from the manufacturer's list-price sheet or catalog.) Assume that a recessed fixture and a corner closet take up 9 ft^2 of ceiling area in the room. L = 12 ft, W = 11 ft, A = 9 ft^2 and U = 444 ft.

$$\text{wire spacing} = \frac{12(12 - 1)(11 - 1) - 12 \times 9}{444}$$

$$= \frac{12 \times 11 \times 10 - 108}{444} = \frac{1320 - 108}{444}$$

$$= \frac{1212}{444} = 2.73 \text{ in. (or } 2\tfrac{3}{4} \text{ in.)}$$

Gypsum wallboard ceiling installation. In addition to the requirements of Fig. 7-2, it is necessary to determine the required spacing between the heating wire runs. This can be done using the same

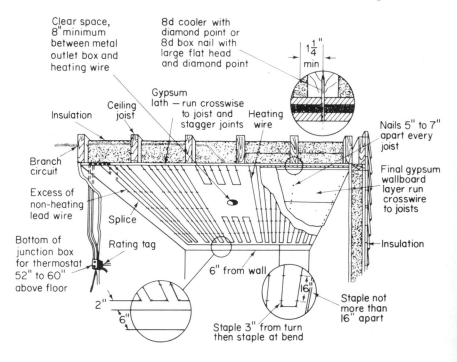

FIGURE 7-2   Basic spacing requirements for radiant heating wire in wallboard ceilings (*Courtesy* General Electric Company).

equation as for plaster ceilings, except that the value of A is not the same. With wallboard ceilings, the value of A = the total square feet of unusable ceiling area that must be allowed for in nailing facing wallboard, plus the other unusable areas, such as those required for cabinets, closets, and lighting fixtures. Because of the nails in joists for wallboard installations, no heating wire run should be installed under a ceiling joist. The unusable area that must be allowed for nailing purposes is determined as follows.

About 3 in. under each ceiling joist should be kept free of heating wire runs. Thus a strip $\frac{1}{4}$ ft wide under each joist run must be included in the unusable area. To determine the unusable area, multiply the length of the joist by the number of joists and by $\frac{1}{4}$ ft.

Note: Never install heating wire in a laminated wallboard ceiling with a wire spacing less than 1 $\frac{15}{16}$ in.

Example of wallboard ceiling wire spacing

The room is the same as for the plaster ceiling example except that the room has 8 joists 12 ft long. L = 12 ft, W = 11 ft, A = 9 ft^2, U = 444 ft, and unusable area for joists = 8 $\times$ 12 $\times$ $\frac{1}{4}$ = 24 ft^2.

Total unusable area = 24 + 9 = 33 ft^2.

$$\text{wire spacing} = \frac{12(12-1)\,(11-1) - 12 \times 33}{444}$$

$$= \frac{12 \times 11 \times 10 - 396}{444} = \frac{1320 - 396}{444}$$

$$= \frac{924}{444} = 2.08 \text{ in. (or } 2\,\tfrac{1}{16} \text{ in.)}$$

Concrete floor installation. The use of heating wire in concrete floors for *complete house heating is not recommended.* The heating contractor should consider several important factors common to all residential radiant heating floor installations before making a decision to install a system.

1. Concrete slabs of the thickness required for floors in residences store up a considerable quantity of heat. Therefore, the slab continues to give off heat when it is not needed, and it is slow to react when heat is required. This is particularly objectionable during the spring and fall months, when night and daytime heating requirements vary considerably. The thermal inertia can be minimized by the installation of as thin a con-

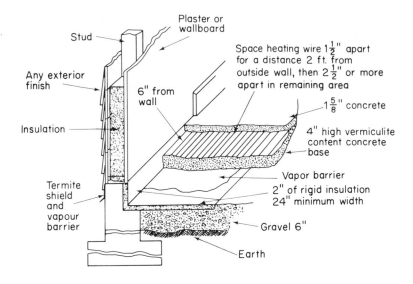

Plaster or wallboard

Stud

Any exterior finish

Insulation

6" from wall

Space heating wire $1\frac{1}{2}$" apart for a distance 2 ft. from outside wall, then $2\frac{1}{2}$" or more apart in remaining area

$1\frac{5}{8}$" concrete

4" high vermiculite content concrete base

Vapor barrier

Termite shield and vapour barrier

2" of rigid insulation 24" minimum width

Gravel 6"

Earth

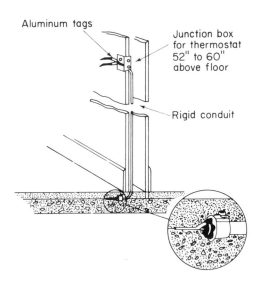

Aluminum tags

Junction box for thermostat 52" to 60" above floor

Rigid conduit

FIGURE 7-3   Basic spacing requirements for radiant heating wire in slab floors (*Courtesy* General Electric Company).

crete slab as is practical. This generally introduces other considerations, however.

2. Experience shows that most people object to high floor

temperatures and that floor temperatures above 85° Fahrenheit are considered uncomfortable. This limitation must be taken into account when planning the system to meet the heating requirements of the room or rooms.

3. Greater care must be exercised in insulating and vapor sealing the slab if a radiant heating floor system is planned.

4. In some cases radiant heating floor systems may be desirable as an *auxiliary* source of heat.

If concrete slab floor heating is installed, use the spacing shown for plaster ceilings, with the following exceptions:

The first run of the heating wire should be installed parallel to the exterior wall 6 in. from the wall. (If the room has more than one exterior wall, the initial run should be installed parallel to the wall that has the greatest heat loss.)

In the area 2 ft from the wall, parallel runs should be spaced $1\frac{1}{2}$ in. apart.

In the remaining floor area, parallel runs should be spaced $2\frac{1}{2}$ in. apart, more as required. In this area *never use a spacing of less than $2\frac{1}{2}$ in.* Closer spacings in this area may produce uncomfortable floor temperatures.

Turns or loops should be made at least 6 in. from the walls. Before making the last turn or loop, install the nonheating power lead up and through the conduit to the thermostat outlet box, as shown in Fig. 7-3. Be sure to provide at least 6 in. of excess for ease in making electrical connections in the outlet box.

7-1.2 Wall and baseboard heaters

Wall and baseboard heaters can be used as the sole heating system or as auxiliary units where supplementary heat is required. Such heaters can be installed to give highly efficient performance in new or old homes, additions, garages, or other locations where heat is desired, provided that proper insulation has been installed.

From a design standpoint, the main concern with wall or baseboard heaters is that they will deliver sufficient heat for the area, type of construction, weather conditions, and so on. As discussed in Sections 7-5 and 7-6, the heat requirements are reduced to a given wattage for a given heating requirement. Thus, if the wall and baseboard heater is of sufficient wattage, it should produce adequate heat.

7-1.3 Utility and ceiling heaters

Utility and ceiling heaters, including bathroom ceiling heaters with exhaust fans and infrared lamp ceiling heaters, are used only in local areas where supplementary heat is required. About the only concern from a design standpoint is that the conductors and overcurrent devices to such heaters are of the correct ampacity.

7-1.4 Thermostats

Perhaps the most important advantage obtained from the use of electric heat is that of room-by-room control. With a thermostat in every room, careful consideration must be given to the selection and placement of the thermostat to ensure the proper operation of the electric heating system.

Thermostats should be located 52 to 60 in. above the floor or inside walls in locations that are not subject to abnormal temperatures. For example, thermostats should not be mounted near doors or windows, on cold walls, or adjacent to cold walls or drafts. Neither should thermostats be mounted near heat sources such as ranges, refrigerators, televisions, lamps, or hot water pipes, or where exposed to the direct rays of the sun. Avoid mounting the thermostat on a wall that has a heat-producing source on the adjacent or reverse side. For example, the thermostat should not be mounted on a dining room wall that is backed up by a kitchen oven.

To provide optimum comfort level control, a thermostat must be able to sense the *average room temperature*. For this reason, a wall-mounted thermostat is recommended. Thermostats built into the various baseboard or wall heater units are subjected to heat from these units and can cause wide variances in room temperature.

Thermostats can be of the line-voltage type, low-voltage type, or combined switch thermostats.

Line-voltage thermostats should be of the type designed especially for electric heating. Their high electrical rating provides for direct control of heating units without the need of relays.

Line-voltage thermostats can be combined under a common wall plate with switches, outlets, and/or pilot lights to provide for stylized control of heating and lighting. Such an arrangement is shown in Fig. 7-4, together with a typical wiring diagram. The thermostat built into the air conditioner is used for control of the air-conditioning unit.

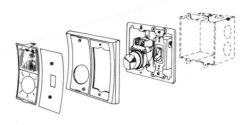

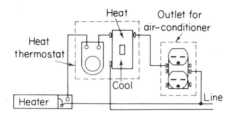

FIGURE 7-4   Line voltage thermostats for electrical heating elements (*Courtesy* General Electric Company).

Low-voltage-control thermostats provide accurate control of both heating and cooling equipment. Figure 7-5 shows typical wiring. Low-voltage-control systems require transformers and relays, as described in Chapter 5.

For heating, most low-voltage thermostats use an adjustable *heat anticipator*, which predicts when the desired temperature will be reached *before* it is reached. This allows the heating system to "coast" to the desired temperature without overshoot. The use of an adjustable anticipator allows the selection of the most desirable cycling rate to provide maximum comfort.

7-1.5 Humidity control

The lack of infiltration in a well-insulated electrically heated home requires that artificial methods be used to control the humidity level. The normal tendency of an electrically heated home is toward high levels of humidity rather than dryness.

Automatic control of exhaust vent fans can be used to maintain a low relative humidity during the heating season. This is done by exhausting the moist inside air and allowing the dryer outside air to infiltrate and replace it.

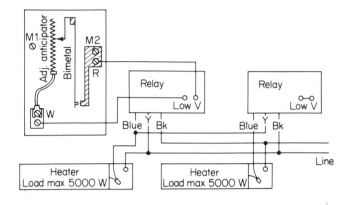

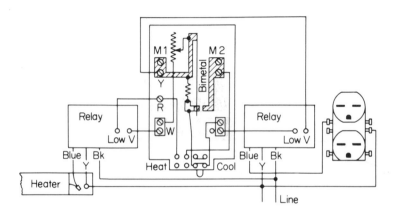

FIGURE 7-5   **Low voltage thermostats for electrical heating elements**
(*Courtesy* **General Electric Company**).

The humidity control, or humidistat, as it is sometimes called, turns the fan on and off at certain humidity levels. The operating element of the humidistat is composed of many strands of carefully selected, specially treated human hair. This element is so sensitive that a slight change in humidity will cause the control to operate. A tension device prevents the hair strands from being damaged by extreme humidity or mechanical strain.

As in the case of thermostats, humidity controls should be located on an inside wall of the room. Do not put the humidistat over or near the range, sink, or in other places of extreme moisture, such

as bathrooms.   Locate the humidity control where natural air cir-
culation is unrestricted.   Also, humidistats should not be installed
where operation might be affected by lamps, sunlight, fireplaces,
registers, radiators, radios, concealed pipes, or room occupants.

7-2 TEMPERATURE-HUMIDITY INDEX

The graph of Fig. 7-6 indicates zones of varying degrees of discomfort, based on the Discomfort Index developed by Earl C. Thom,
Meterologist of the U.S. Weather Bureau. The dashed line indicates
the combination of humidity and temperature at which most people
are comfortable. For example, if the relative humidity is 30 per cent,
most people will be comfortable at 80°F. If the relative humidity is
increased to 60 per cent, the temperature must be reduced to about
75°F, for comfort.

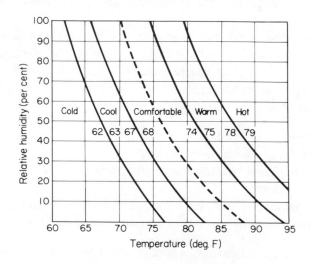

FIGURE 7-6   Temperature humidity index (*Courtesy* General Electric
Company).

Above the humidity and temperature combinations shown by the
dashed line, increasing numbers of people are uncomfortably warm.
Below the dashed line, increasing numbers of people experience a
feeling of cold.   In the zone labeled "warm," 90 per cent of the
people are uncomfortable, and in the zone labeled "hot," everyone
is uncomfortable.

7-3 NEED FOR INSULATION

In order to maintain a room or home at a comfortable temperature, heat must be supplied in an amount equal to that lost due to (1) *transmission losses*, or the heat lost by conduction through walls, windows, doors, ceilings, floors, and so on, and (2) *infiltration losses*, the heat required to warm cold air that filters through cracks and openings, such as when doors are opened. (The cold air displaces the warm air in the room lost to the outdoors.)

7-3.1 Proper insulation for transmission losses

Insulation should be installed in accordance with local electric utility and/or electric heating code specifications. To assure an efficient and economical electrical heating installation, insulation must meet or exceed NEMA (National Electrical Manufacturers Association), FHA (Federal Housing Authority), and industry requirements.

Although the exact requirements for insulation will vary with the type of construction and the region where the building is located, a well-insulated house or room requires less heat than a poorly insulated house. In most regions the overall economics of initial insulation cost versus electric heating costs justify:

1. Six inches of insulation in the ceiling, $3\frac{5}{8}$ in. in the side walls, and 2 in. in floors over unheated areas.
2. Double glazed windows or tight-fitting storm windows.
3. Storm doors.
4. The installation of at least 2 in. of insulation in ceilings between floors in multistory homes and apartment houses.

When these specifications are met, the *heat loss* will generally be lower than industry requirements.

7-3.2 Minimizing air leakage to reduce infiltration losses

Care should be exercised in construction to minimize infiltration. Weatherstripping of windows and caulking around window frames and exterior door frames reduces area (crackage) through which cold air can enter the home or room.

Tight-fitting storm windows and storm doors help reduce infiltration. Fireplaces should be equipped with tight-fitting dampers.

7-3.3 Construction pointers for low heat loss

The following recommendations should not be overlooked in providing for low heat loss.

Living areas. Increased moisture in insulation decreases its insulating value. Therefore, vapor barriers should be installed to prevent a loss of insulation efficiency. The vapor barrier can be installed as an integral part of the insulation, or as a separate membrane, such as a 0.004-in. polyethylene sheet. The vapor barrier should be applied so that it is on the warm side of the insulation in walls, ceilings, and floors over unheated areas. The barrier should have a moisture-vapor transmission rate of 1 *perm* (refer to Section 7-4) or less.

In lieu of, or in addition to vapor barriers, humidistat-controlled ventilating fans are highly recommended. If blown insulation is used, it should be applied according to the rated density of the materials used.

A small exhaust fan should be installed in each bathroom area, and at least one humidistat-controlled exhaust fan of not less than 150 to 200 cubic feet per minute (ft^3/min) capacity should be installed in the kitchen or laundry area.

A fan capacity of 450 ft^3/min is recommended for existing homes where vapor barriers cannot be installed. A humidistat setting of 25 to 35 per cent relative humidity is recommended for the heating season.

To eliminate possible moisture problems, the clothes dryer must be exhausted to the outside not to the attic or other enclosed areas.

Attic or space between top floor ceiling and roof. Cross ventilation should be provided for each enclosed area by a minimum of two louvers or similar devices that adequately protect against snow and rain. The total net openings (actual ventilating areas) of two louvers should be $\frac{1}{300}$, or 0.0033, of the enclosed ceiling area when a *vapor barrier is used. If a vapor barrier is not used,* the total net areas of the openings (louvers) must be doubled (at least $\frac{1}{150}$, or 0.0066, of the enclosed ceiling area).

In attics that can be converted into usable living space or storage area, 50 per cent of the required ventilating area should be provided near the peak of the roof, above the height of the future ceiling. Ventilation should be provided *above all insulation in ceilings.*

Basements. Walls of finished rooms in basements, such as recreation rooms, should be insulated if minimum heating costs and good comfort conditions are desired.

Crawl spaces. A 0.004-in polyethylene sheet or other suitable permanent vapor barrier should be placed over all the ground in crawl spaces. To prove good protection, joints must lap 4 to 6 in. The crawl-space area must be vented by at least two wall ventilators. Their total net areas should not be less than $\frac{1}{1500}$, or 0.00067, of the crawl-space area. *If a vapor barrier is not used*, four foundation wall ventilators must be provided, one near each of the four corners. Their total net ventilating openings should not be less than $\frac{1}{150}$, of the crawl-space area.

Concrete slabs.   All the gravel should be covered with a vapor barrier such as 0.006-in. polyethylene or other suitable permanent material. The sheets should overlay 4 to 6 in. with all joints properly sealed. The gravel fill should be at least 4 in. thick and the slab laid several inches above grade with effective subsoil drainage. Two inches of insulation should be installed between the edge of the slab and the foundation.

If supplementary heating is installed in the slab, 2 in. of insulation should be installed over the gravel adjacent to the inside of the foundation wall. The insulation should extend 18 to 24 in. under the slab.

7-4 HEATING DEFINITIONS

Before going into the details of heating calculations, we shall define some of the terms used by the heating industry.

Temperature. Temperature is defined as the rate of movement of molecules and therefore indicates the speed of movement of the particles of every substance, including air.

Heat. Heat is defined as the energy due to this rate of movement of molecules and is therefore dependent on the speed of the molecule (temperature) and the number of molecules involved (mass or cubical content).

Infiltration. Infiltration is the heat that is lost through cracks around doors and windows and normal use of doors. Infiltration is measured by *air changes per hour*. A rate of one-half or three-quarters of one air change per hour is suggested for average construction. The frequent use of doors in commercial buildings increases the infiltration rate.

Btu. The *British thermal unit* (Btu) is a standard unit of heat, the heat required to raise 1 lb of water $1°F$. Note that Btu involves quantity (1 lb of water) and temperature ($1°F$ rise). Thus Btus give the heat involved and not the rate of flow of heat. 1 Btu/hr = 0.293 W.

Watts. Wattage is defined as the rate of flow of electrical energy (not the quantity, the rate). Since electrical energy can be converted into heat energy without any loss, watts can be considered the rate of flow of heat (similar to a flow of so many gallons of water per hour passing through a pipe.) 1000 W = 1 kW = 3412 Btu/hr.

Watthours. Watthours are a quantity of electrical energy, a product of a rate of flow times the length of time. One watthour = 3.41 Btus.

Kilowatthours. One kilowatthour = 1000 watthours.

Outdoor Design Temperature. The *outdoor design temperature* is the outdoor temperature (OT) used in designing a heating system for a particular locality. OT, which is generally accepted as being $15°$ more than the lowest temperature on record, is usually available from the U.S. Weather Bureau or the utility company.

Degree-days. *Degree-days* (DD) is a relative measure of the heat requirement for an average year for any locality. Degree-days in a heating season equal the total of degree-days that have occurred on each day in the heating season. The degree-day units in 1 day are obtained by subtracting the mean temperature for the day from $65°F$. For example, if the highest temperature recorded during a day is $40°F$ and the lowest is $10°F$, the mean temperature is $25°F$, or $(40 + 10)/2 =$ 25. This would be recorded as 40 degree-day units, or $65 - 25 = 40$. Degree-days are calculated each day by local weather bureaus. Daily, monthly, and seasonal totals are printed in many newspapers.

Temperature Differential. The *temperature differential* (TD) is the difference between the outdoor design temperature and $70°F$, the standard indoor design temperature for electric heating. For example, if the outdoor design temperature is $25°F$, then the TD is $45°F$. If the outdoor design temperature is $-15°F$, then the TD is $85°F$.

U factor. The *U factor* is the thermal transmittance or conductance of a substance, that is, the overall coefficient of the heat transfer through a section of the wall, floor, insulation, or whatever. The U factor is equal to the amount of heat in Btus transferred through a 1-ft^2 section of the wall of the building in 1 h when the difference in temperature of the two surfaces is $1°F$. The *insulating value* of any insulating material is expressed as the K factor of the material.

W factor. The *W factor* is similar to the U factor but converted into electrical terms for simplified calculation in electric heating in-

heat loss factors

Description	Watts/sq.ft. Per °F TD	90 TD –20 OT	80 TD –10 OT	70 TD 0 OT	60 TD 10 OT	50 TD 20 OT	40 TD 30 OT	40 TD 40 OT
LOSS THROUGH SOLID BRICK WALLS								
WALL THICKNESS—8 INCHES								
Plain brick, no inside finish	0.147	13.23	11.76	10.29	8.82	7.35	5.88	4.41
Inside plaster direct on walls (no lath), no furring	0.135	12.15	10.80	9.45	8.10	6.75	5.40	4.05
Inside furred, with plaster	0.088	7.92	7.04	6.16	5.28	4.40	3.52	2.64
Inside 1/2" rigid insulation furred on brick	0.064	5.76	5.12	4.48	3.84	3.20	2.56	1.92
Inside furred, with 1" insulation, and plaster	0.041	3.69	3.28	2.87	2.46	2.05	1.64	1.23
Inside furred, with 2" insulation, and plaster	0.029	2.61	2.32	2.03	1.74	1.45	1.16	0.87
Inside furred, with 3 5/8" insulation, and plaster	0.022	1.98	1.76	1.54	1.32	1.10	0.88	0.66
WALL THICKNESS—12 INCHES								
Plain brick, no inside finish	0.105	9.45	8.40	7.35	6.30	5.25	4.20	3.15
Inside plaster direct on walls (no lath), no furring	0.100	9.00	8.00	7.00	6.00	5.00	4.00	3.00
Inside furred with plaster	0.070	6.30	5.60	4.90	4.20	3.50	2.80	2.10
Inside 1/2" rigid insulation furred on brick	0.056	5.04	4.48	3.92	3.36	2.80	2.24	1.68
Inside furred, with 1" insulation, and plaster	0.038	3.42	3.04	2.66	2.28	1.90	1.52	1.14
Inside furred, with 2" insulation, and plaster	0.027	2.43	2.16	1.89	1.62	1.35	1.08	0.81
Inside furred, with 3 5/8" insulation, and plaster	0.021	1.89	1.68	1.47	1.26	1.05	0.84	0.63
WALL THICKNESS—16 INCHES								
Plain brick, no inside finish	0.082	7.38	6.56	5.74	4.92	4.10	3.28	2.46
Inside plaster direct on walls (no lath), no furring	0.079	7.11	6.32	5.53	4.74	3.95	3.16	3.37
Inside furred, with plaster	0.059	5.31	4.72	4.13	3.54	2.95	2.36	1.77
Inside 1/2" rigid insulation furred on brick	0.050	4.50	4.00	3.50	3.00	2.50	2.00	1.50
Inside furred, with 1" insulation, and plaster	0.035	3.15	2.80	2.45	2.10	1.75	1.40	1.05
Inside furred, with 2" insulation, and plaster	0.028	2.52	2.24	1.96	1.68	1.40	1.12	0.84
Inside furred, with 3 5/8" insulation, and plaster	0.019	1.71	1.52	1.33	1.14	0.95	0.76	0.57
LOSS THROUGH BRICK VENEER WALLS (wood framing and sheathing)								
Inside lath and plaster, no insulation	0.079	7.11	6.32	5.53	4.74	3.95	3.16	2.37
Inside 1/2" insulating board	0.062	5.58	4.96	4.34	3.72	3.10	2.48	1.86
Inside plaster, with 1" insulation	0.041	3.69	3.28	2.87	2.46	2.05	1.64	1.23
Inside plaster, with 2" insulation	0.028	2.52	2.24	1.96	1.68	1.40	1.12	0.84
Inside plaster, with 3 5/8" insulation	0.021	1.89	1.68	1.47	1.26	1.05	0.84	0.63
LOSS THROUGH CONCRETE BLOCK WALLS (8" concrete block with air cells, gravel aggregate)								
No interior finish, no insulation	0.164	14.76	13.12	11.48	9.84	8.20	6.56	4.92
Inside plaster on block (no lath), no insulation	0.152	13.68	12.16	10.64	9.12	7.60	6.08	4.56
Inside furred, with plaster	0.094	8.46	7.52	6.58	5.64	4.70	3.76	2.82
Inside furred, with plaster, and 1" insulation	0.044	3.96	3.52	3.08	2.64	2.20	1.76	1.32
Inside furred, with plaster, and 2" insulation	0.029	2.61	2.32	2.03	1.74	1.45	1.16	0.87
Inside furred, with plaster, and 3 5/8" insulation	0.022	1.98	1.76	1.54	1.32	1.10	0.88	0.66

FIGURE 7-7(a) Heat loss factors for walls and ceilings, sheet 1 (*Courtesy* **General Electric Company**).

heat loss factors — continued

	Watts/sq ft Per °F TD	90 TD -20 OT	80 TD -10 OT	70 TD 0 OT	60 TD 10 OT	50 TD 20 OT	40 TD 30 OT	30 TD 40 OT
LOSS THROUGH BRICK AND CONCRETE BLOCK WALLS (4" brick and 8" concrete block)								
No interior finish, no insulation	0.129	11.61	10.32	9.03	7.74	6.45	5.16	3.87
Inside plaster on block (no lath), no insulation	0.120	10.80	9.60	8.40	7.20	6.00	4.00	3.60
Inside furred, with plaster	0.082	7.38	6.56	5.74	4.92	4.10	3.28	2.46
Inside furred, with plaster, and 1" insulation	0.041	3.69	3.28	2.97	2.46	2.05	1.64	1.23
Inside furred, with plaster, and 2" insulation	0.029	2.61	2.32	2.03	1.74	1.45	1.16	0.87
Inside furred, with plaster, and 35/8" insulation	0.022	1.98	1.76	1.54	1.32	1.10	0.88	0.66
LOSS THROUGH FRAME WALLS WITH STUCCO EXTERIOR (1" wood sheathing and moisture barrier)								
Inside plaster, no insulation	0.094	8.46	7.52	6.58	5.64	4.70	3.76	2.82
Inside 1/2" insulating board	0.064	5.76	5.12	4.48	3.84	3.20	2.56	1.92
Inside plaster, with 1" insulation	0.041	3.69	3.28	2.87	2.46	2.05	1.64	1.23
Inside plaster, with 2" insulation	0.029	2.61	2.32	2.03	1.74	1.45	1.16	0.87
Inside plaster, with 35/8" insulation	0.022	1.98	1.76	1.54	1.32	1.10	0.88	0.66
LOSS THROUGH FRAME WALLS WITH WOOD SIDING OR SHINGLES (1" wood sheathing and moisture barrier)								
Inside plaster	0.076	6.84	6.08	5.32	4.56	3.80	3.04	2.28
Inside 1/2" insulating board	0.055	4.95	4.40	3.85	3.30	2.75	2.20	1.65
Inside 1/2" insulating board and plaster (no lath)	0.055	4.95	4.40	3.85	3.30	2.75	2.20	1.65
Inside 1" insulating board and plaster (no lath)	0.044	3.96	3.52	3.08	2.64	2.20	1.76	1.32
Inside plaster, with 1" insulation	0.032	2.88	2.56	2.24	1.92	1.60	1.28	0.96
Inside plaster, with 2" insulation	0.025	2.25	2.00	1.75	1.50	1.25	1.00	0.75
Inside plaster, with 3 5/8" insulation	0.020	1.80	1.60	1.40	1.20	1.00	0.90	0.60
LOSS THROUGH INTERIOR WALLS								
Studding with plaster on one side	0.178	16.02	14.24	12.46	10.68	8.90	7.12	5.34
Studding with plaster on both sides	0.100	9.00	8.00	7.00	6.00	5.00	4.00	3.00
Brick plastered on both sides (no lath)	0.126	11.34	10.08	8.82	7.56	6.30	5.04	3.78
Studding with 1/2" insulating board, one side	0.105	9.45	8.40	7.35	6.30	5.25	4.20	3.15
Studding with 1/2" insulating board, two sides	0.056	5.04	4.48	3.92	3.36	2.80	2.24	1.68
Studding with 1" insulation	0.044	3.96	3.52	3.08	2.64	2.20	1.76	1.32
Studding with 2" insulation	0.025	2.25	2.00	1.75	1.50	1.25	1.00	0.75
Studding with 3" insulation	0.023	2.07	1.84	1.61	1.38	1.15	0.92	0.69
LOSS THROUGH FLOORS (over air space)								
6" concrete, bare	0.173	15.57	13.84	12.11	10.38	8.65	6.92	5.19
4" concrete, bare	0.202	18.18	16.16	14.14	12.12	10.10	8.08	6.06
4" concrete with asphalt tile	0.173	15.57	13.84	12.11	10.38	8.65	6.92	5.19
4" concrete with double floor on sleepers	0.073	6.57	5.84	5.11	4.38	3.65	2.92	2.19
Double wood floor over ground, solid and unventilated foundation	0.100	9.00	8.00	7.00	6.00	5.00	4.00	3.00
Double wood floor with plaster beneath	0.070	6.30	5.60	4.90	4.20	3.50	2.80	2.10
Double wood floor with 1/2" insulation board beneath joists	0.056	5.04	4.48	3.92	3.36	2.80	2.24	1.68
Double wood floor over ground with asphalt tile	0.088	7.92	7.04	6.16	5.28	4.40	3.52	2.64
Double wood floor over ground with 1" insulation	0.044	3.96	3.52	3.08	2.64	2.20	1.76	1.32
Double wood floor over ground with 2" insulation	0.029	2.61	2.32	2.03	1.74	1.45	1.16	0.87
Double wood floor over ground with 3" insulation	0.021	1.89	1.68	1.47	1.26	1.05	0.84	0.63

FIGURE 7-7(b)   Heat loss factors for walls and ceilings, sheet 2   (*Courtesy* General Electric Company).

LOSS THROUGH CEILINGS

	Watts/sq.ft. Per °F TD	90 TD -20 OT	80 TD -10 OT	70 TD 0 OT	60 TD 10 OT	50 TD 20 OT	40 TD 30 OT	30 TD 40 OT
Single wood under joists, no flooring above	0.132	11.88	10.56	9.24	7.82	6.60	5.28	3.96
Plaster, no flooring above	0.179	16.11	14.32	12.53	10.74	8.95	7.16	5.37
Plaster with double flooring above	0.070	6.30	5.60	4.90	4.92	3.50	2.80	2.10
Plaster with 2" insulation and double flooring above	0.026	2.34	2.08	1.82	1.56	1.30	1.04	0.78
Plaster with 1" insulation above	0.056	5.04	4.48	3.92	3.36	2.80	2.24	1.68
Plaster with 2" insulation above	0.035	3.15	2.80	2.45	2.10	1.75	1.40	1.05
Plaster with 3" insulation above	0.023	2.07	1.84	1.61	1.38	1.15	0.92	0.69
Plaster with 6" insulation above	0.014	1.26	1.12	0.98	0.84	0.70	0.56	0.42
1/2" gypsum board with 3" insulation above	0.021	1.89	1.68	1.47	1.26	1.05	0.84	0.63
1/2" gypsum board with 6" insulation above	0.014	1.26	1.12	0.98	0.84	0.70	0.56	0.42

LOSS THROUGH WINDOWS, DOORS, AND GLASS

	Watts/sq.ft. Per °F TD	90 TD -20 OT	80 TD -10 OT	70 TD 0 OT	60 TD 10 OT	50 TD 20 OT	40 TD 30 OT	30 TD 40 OT
Single glass	0.331	29.79	26.48	23.17	19.86	16.55	13.24	9.93
Single glass with storm window	0.155	13.95	12.40	10.85	9.30	7.75	6.20	4.65
Double glass (3/4" air space)	0.158	14.22	12.64	11.06	9.48	7.90	6.32	4.74
Double glass (1/4" air space)	0.179	16.11	14.32	12.53	10.74	8.95	7.16	5.37
Skylight single glass	0.410	36.90	32.80	28.70	24.60	20.50	16.40	12.30
Skylight double glass (3/4" air space)	0.187	16.83	14.96	13.09	11.22	9.35	7.48	5.61
Hollow glass block wall (7 3/4" x 7 3/4" x 3 7/8" thick with glass fiber screen dividing the cavity)	0.141	12.69	11.28	9.87	8.46	7.05	5.64	4.23
Solid wood door, exposed to outside (1 3/8" actual thickness)	0.149	13.41	11.92	10.43	8.94	7.45	5.96	4.47
Solid wood door (1 3/8" actual thickness) with glass storm door	0.088	7.92	7.04	6.16	5.28	4.40	3.52	2.64

LOSS THROUGH ROOFS

	Watts/sq.ft. Per °F TD	90 TD -20 OT	80 TD -10 OT	70 TD 0 OT	60 TD 10 OT	50 TD 20 OT	40 TD 30 OT	30 TD 40 OT
Flat metal roof, no insulation beneath	0.275	24.75	22.00	19.25	16.50	13.75	11.00	8.25
Slate or tile on sheathing, no insulation	0.161	14.49	12.88	11.27	9.66	8.05	6.44	4.83
Asphalt shingles or wood roofing, no insulation	0.155	13.95	12.44	10.85	9.30	7.75	6.20	4.65
Wood shingles, no insulation	0.141	12.69	11.28	9.87	8.46	7.05	5.64	4.23
Flat with tar and gravel, tar and gravel board	0.088	7.92	7.04	6.16	5.28	4.40	3.52	2.64
Asphalt shingles with 1/2" insulating board	0.067	6.03	5.36	4.69	4.02	3.35	2.68	2.01
Slate or tile with 1/2" insulating board	0.070	6.30	5.60	4.90	4.20	3.50	2.80	2.10
Wood shingles with 1/2" insulating board	0.064	5.76	5.12	4.48	3.84	3.20	2.56	1.92
Asphalt shingles with 3" insulation	0.025	2.25	2.00	1.75	1.50	1.25	1.00	0.75
Slate or tile with 3" insulation	0.025	2.25	2.00	1.75	1.50	1.25	1.00	0.75
Wood shingles with 3" insulation	0.024	2.16	1.92	1.68	1.44	1.20	0.96	0.72

INFILTRATION FACTOR

	Watts per cu. ft./°F	90 TD -20 OT	80 TD -10 OT	70 TD 0 OT	60 TD 10 OT	50 TD 20 OT	40 TD 30 OT	30 TD 40 OT
	0.0053	0.48	0.42	0.37	0.32	0.27	0.21	0.16

FIGURE 7-7(c)   Heat loss factors for walls and ceilings, sheet 3   (*Courtesy* General Electric Company).

stallations. The W factor is stated in watts per square foot per degree temperature difference. The W factor is the coefficient found in the table of Fig. 7-7 and is used to find the multiplying factors given under the various TD and OT values in Fig. 7-7.

Perms. *Perms* are defined as the rate of flow of water vapor through any substance, usually a vapor barrier. Consequently, perms is a measure of the quality of any particular vapor barrier. One perm is equal to a flow of 1 grain of water per square foot per hour per inch of mercury vapor-pressure difference.

R number. The *R number* indicates a performance standard of insulation installed and is based on the thermal resistance of the material. The R numbers that meet the requirements of the tables in Fig. 7-7 are R-4, 1-in. insulation; R-7 2-in. insulation; R-11, 3-in. insulation; R-13, $3\frac{5}{8}$-in. insulation; and R-22, 6-in. insulation.

7-5 STANDARD METHOD OF CALCULATING HEAT LOSS

The Electric House Heating Section of the National Electrical Manufacturers Association (NEMA) has issued a standard procedure for the calculation of heat losses or heating requirements. It is recommended that the heating requirements be determined by this method *before* accepting any job or beginning any installation work.

The following method takes into consideration outdoor design temperatures, a 70°F indoor design temperature condition, *annual* degree days, and heat-loss factors for various types of construction.

The tables of Fig. 7-7 contain heat-loss factors in *watts per square feet* for various types of construction and parts of a home at different outdoor design temperatures. The watts-per-square-foot figures are based on an indoor temperature of 70°F.

The table of Fig. 7-8 provides heat-loss factors for slab construction in *watts per linear feet* of exposed edge for different outdoor design temperatures.

The table of Fig. 7-9 provides heat-loss factors in *watts per square feet* for floors and walls below grade in different cities.

The information in Figs. 7-8 and 7-9 is based on a 70°F indoor temperature.

The temperature differential (the difference between the outdoor and indoor, or 70°F, design temperature conditions) affects the heat-loss factor. For example, in a –10° area, the TD is 80, whereas is a +10° area, the TD is 60, for 70°F indoor conditions.

	heat loss for slab construction					
Heat losses are in WATTS PER LINEAR FOOT OF EXPOSED EDGE AT DESIGN CONDITIONS AND WITH EDGE INSULATION AS INDICATED. Factors for "no insulation" and one inch thick vertical insulation are shown for comparison purposes. Neither are recommended for heating installations.	WATTS LOSS PER LINEAR FOOT OF EXPOSED EDGE					
OUTDOOR DESIGN TEMPERATURE ⟶	–20 F	–10 F	0 F	10 F	20 F	30 F
EDGE INSULATION						
No Insulation—Edge Exposed	22	19	17	14	12	9
1" Thick Vertical Extending Down 18" Below Floor Surface	16	14.5	13.2	11.7	9.6	8
2" Thick Vertical Extending Down 24" or 2" Thick L-Type Extending Down 12" and Under 12" or more	14.6	13.2	11.7	10.2	8.8	7.3

FIGURE 7-8   Heat loss factors for slab construction (*Courtesy* General Electric Company).

	heat loss factors for below floors and walls	
	WATTS PER SQ. FT.	
LOCATION	Basement Floors	Below Grade Walls
Augusta, Maine; Buffalo, N.Y.; Flint, Michigan; St. Paul, Minn; Helena, Montana	0.75	1.5
Boston, Springfield, Mass; Pittsburgh, Detroit, Chicago, Des Moines, Denver, Sante Fe, Salt Lake City	0.6	1.2
New York, Philadelphia, Cincinnati, Youngstown, Springfield, Illinois; Kansas City, Portland, Seattle	0.45	0.9
Washington, D.C.; Norfolk, Winston-Salem, Knoxville, Louisville, Chattanooga, Tulsa, Amarillo, San Francisco	0.3	0.6
Raleigh, Wilmington, N.C.; Columbia, S.C.; Atlanta, Birmingham, Memphis, Little Rock, Oklahoma City, Fort Worth, El Paso, Phoenix, Sacramento	0.15	0.3
Jacksonville, New Orleans, Houston, Los Angeles	0.0	0.0

FIGURE 7-9   Heat loss factors for below grade floors and walls (*Courtesy* General Electric Company).

The annual degree-days for a city can be obtained from the local U.S. Weather Bureau or the local utility company. Annual degree-days have been used for many years to establish the heating needs for homes in given areas.

Since electric heating is installed to provide comfort conditions in each room, it is essential that the heating requirements of *each room* be established.

7-5.1 Work-sheet calculations

The use of a work sheet or form similar to that of Fig. 7-10 is recommended to facilitate the calculations, and as a permanent record of the requirements for the home. Use the following procedure in conjunction with the form.

1. Fill in the total area of windows and exterior (exposed) doors in square feet. Doors and windows are grouped, as most doors have some glass area and therefore have comparable factors.
2. Insert the total net area of the exterior wall or walls. This equals the total wall area in square feet minus the total area of windows and exterior doors in the room.
3. Fill in the total area of the floor in square feet and that of the ceiling in square feet. For floors on a slab, use the linear footage of the exposed edge of the slab.
4. Determine the cubic size of the room by multiplying the floor area in square feet by the height of the room in feet.
5. Having filled in the required information for each room, obtain the proper heat-loss factors from the tables in Figs. 7-7 through 7-9. Use Fig. 7-7 for everything but slab floors (Fig. 7-8) and below-grade floors and walls (Fig. 7-9).
6. The heat loss in watts for each part of the room is obtained by multiplying the two numbers (such as area in square feet by heat-loss factor).
7. The infiltration loss for each room is obtained by multiplying the number of cubic feet in the room by the infiltration factor and by the number of air changes per hour. In most areas, one air change per hour is an accepted figure. Thus (cubic feet $\times$ infiltration factor $\times$ 1) = infiltration loss per room where one air change is accepted practice. Constant exhaust in the kitchen, bath, or lavatory may increase air changes.
8. The total watts lost (heating requirements) per room is obtained by adding the watts lost in each part to the infiltration losses.

7-5.2 Example of work-sheet calculations

The calculations for the home in Fig. 7-11 are shown on the form of Fig. 7-10. Following are two examples. The glass and door for

ELECTRIC RESISTANCE HEATING
HEAT LOSS CALCULATION
For Specification of Electric Heating Systems

Residence of _John Doe_ / _Somewhere_ / _U.S.A._
          Name   City   State

Inside Design Temp.	_70°_	Insulation:
Outside Design Temp.	_0°_	Ceiling _6"_
Degree Days	_5000_	Outside Walls _3 5/8"_
Estimated KWHR Use	_14,300_	Floor _2"_
Power Rate	_$.0151 Kwhr_	Storm Windows & Doors _Yes._
Estimated Annual Cost	_$ 215.00_	Total Heat Loss _10.8 Kw_

INSTALLED CAPACITY 11.6 KW

ROOM	SECTION	FOOTAGE	FACTOR	HEAT LOSS	ROOM	SECTION	FOOTAGE	FACTOR	HEAT LOSS
Bedroom #1	Glass & Door	24 Sq.Ft.	10.85	260	Hall and Closet	Glass & Door	Sq.Ft.		
	Outside Wall	72 Sq.Ft.	1.40	108		Outside Wall	24 Sq.Ft.	1.40	34
	Floor	132 Sq.Ft.	2.03	268		Floor	81 Sq.Ft.	2.03	164
	Ceiling	132 Sq.Ft.	.98	129		Ceiling	81 Sq.Ft.	.98	79
	Infiltration	1056 Cu.Ft.	.37	391		Infiltration	648 Cu.Ft.	.37	240
	Installed KW 1.2		TOTAL	1149		Installed KW		TOTAL	517
Bedroom #2	Glass & Door	30 Sq.Ft.	10.85	326	Living room	Glass & Door	69 Sq.Ft.	10.85	749
	Outside Wall	154 Sq.Ft.	1.40	216		Outside Wall	235 Sq.Ft.	1.40	329
	Floor	132 Sq.Ft.	2.03	268		Floor	390 Sq.Ft.	2.03	792
	Ceiling	132 Sq.Ft.	.98	129		Ceiling	390 Sq.Ft.	.98	382
	Infiltration	1056 Cu.Ft.	.37	391		Infiltration	3120 Cu.Ft.	.37	1152
	Installed KW 1.4		TOTAL	1330		Installed KW 3.8		TOTAL	3406
Bedroom #3	Glass & Door	30 Sq.Ft.	10.85	326	Kitchen	Glass & Door	33 Sq.Ft.	10.85	358
	Outside Wall	194 Sq.Ft.	1.40	272		Outside Wall	143 Sq.Ft.	1.40	200
	Floor	192 Sq.Ft.	2.03	390		Floor	22 Sq.Ft.	13.2	290
	Ceiling	192 Sq.Ft.	.98	188		Ceiling	120 Sq.Ft.	.98	118
	Infiltration	1536 Cu.Ft.	.37	568		Infiltration	960 Cu.Ft.	.37	355
	Installed KW 2.0		TOTAL	1744		Installed KW 1.4		TOTAL	1321
Bath	Glass & Door	3 Sq.Ft.	10.85	33	Lavette	Glass & Door	6 Sq.Ft.	10.85	65
	Outside Wall	61 Sq.Ft.	1.40	85		Outside Wall	122 Sq.Ft.	1.40	171
	Floor	96 Sq.Ft.	2.03	195		Floor	16 Sq.Ft.	13.2	211
	Ceiling	96 Sq.Ft.	.98	94		Ceiling	60 Sq.Ft.	.98	59
	Infiltration	768 Cu.Ft.	.37	284		Infiltration	480 Cu.Ft.	.37	178
	Installed KW 1.0		TOTAL	691		Installed KW 0.8		TOTAL	684
	Glass & Door	Sq.Ft.				Glass & Door	Sq.Ft.		
	Outside Wall	Sq.Ft.				Outside Wall	Sq.Ft.		
	Floor	Sq.Ft.				Flood	Sq.Ft.		
	Ceiling	Sq.Ft.				Ceiling	Sq.Ft.		
	Infiltration	Cu.Ft.				Infiltration	Cu.Ft.		
	Installed KW		TOTAL			Installed KW		TOTAL	

Notes:

Calculated by_____
Date_____

FIGURE 7-10 Electrical heating worksheet (*Courtesy* General Electric Company).

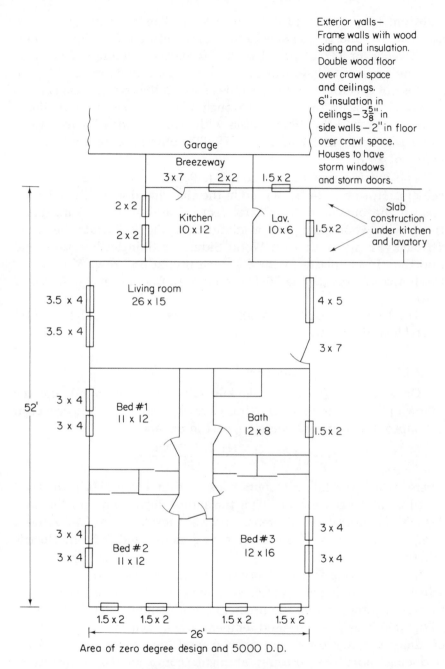

FIGURE 7-11   **Floor plan of house used in example of electric heating calculations** (*Courtesy* **General Electric Company**).

bedroom 1 has 24 ft² (3 × 4, plus 3 × 4). The inside design temperature is 70°, and the outside design temperature is 0°. Thus there is a difference, or TD, of 70°. Use the 70 TD/0 OT column of Fig. 7-7.

The house is to have storm windows and storm doors, but double glass is not specified. Thus it can be assumed that single glass is used. In Fig. 7-7 locate the "Loss Through Windows, Doors, and Glass" section, and find the "Single glass with storm window" row. Move along this horizontal row to the 70 TD column, to find a heat-loss factor of 10.85.

The outside wall of bedroom 1 is 96 ft² (12 ft long by 8 ft of ceiling). However, 24 ft² is used in the glass and doors. Thus the net area of the outside wall is 72 ft² feet. The walls are 3 ⅝ in. thick, frame with wood siding and insulation. In Fig. 7-7 locate the "Loss Through Frame Walls with Wood Siding or Shingles" section, and find the "Inside plaster, with 3 ⅝ -inch insulation" row. Move along this horizontal row to the 70 TD column, to find a heat-loss factor of 1.40.

Total up the heat loss for all rooms as shown in Fig. 7-10 to find a total heat loss of 10.8 kW.

7-5.3 Operating-cost calculations

Once the total heat loss in kilowatts has been found, use the following equation to obtain the estimated annual operating cost.
approximate annual operating cost in dollars

$$= D = \text{kWh} \times R = \frac{\text{HL} \times \text{DD} \times \text{C}}{\text{TD}} \times R$$

where kWh is the kilowatthours consumed in 1 year, HL is the total heat loss of home (in kW), DD is the annual degree-days for the area, TD is the temperature differential between inside design and outdoor design temperatures, C is the correcting factor, and R is the electric rate (in $1kWh).

The correcting factor, C, takes into consideration conditions that affect the operating cost but cannot be factored into the calculations to establish the heating requirements, for example, the living habits of the family as to the desired temperature in each room, the number of times exterior doors are opened, and the use of night setback (reducing room temperatures at night, compared to temperatures during the day). In addition, the actual operating cost will be affected by weather conditions during the year.

From experience with many thousands of electric heating installations, a conservative value for C is 18.5. For carefully operated installations under favorable conditions, the value of C can be assumed to be as low as 12. Under extreme conditions it can go as high as 24.

Experience in any given area will dictate the value to use for C in that area. It is advisable to obtain the local utility company's recommended value for C.

Using the information shown in Fig. 7-10, the estimated annual operating cost (based on a C of 18.5) is $215.00:

$$HL = 10.8 \text{ kW}$$
$$DD = 5000$$
$$TD = 70° (70°F - 0°F)$$
$$C = 18.5 \text{ assumed for this example}$$
$$R = \$0.015/\text{kWh} (1\tfrac{1}{2} \text{ cents/kWh})$$
$$D = \frac{10.8 \times 5000 \times 18.5}{70} \times 0.015 = 215$$

7-6 SHORT METHOD OF CALCULATING HEAT LOSS AND COSTS

It is recommended that the heat loss of each room be calculated according to the standard method described in Section 7-5. However, a short method developed by the General Electric Company is presented in this section. This short method is a quick means of obtaining the heating requirements of a room or home for *estimating purposes*. It is not intended that the short method be used as the sole means of sizing an installation.

7-6.1 Calculating heat loss by room size

Calculating heating loss with the short method involves the use of Figs. 7-12, 7-13, and 7-14.

In Fig. 7-12 the United States is broken into zones. To estimate the heat loss of a room or home in your area, obtain the zone number from Fig. 7-12.

Figure 7-13 contains heat-loss figures for rooms of various sizes in different zones. The heat loss, expressed in watts, applies to rooms having one exterior wall with full insulation and a total glass and door area not to exceed 20 per cent of the total area of the exterior wall.

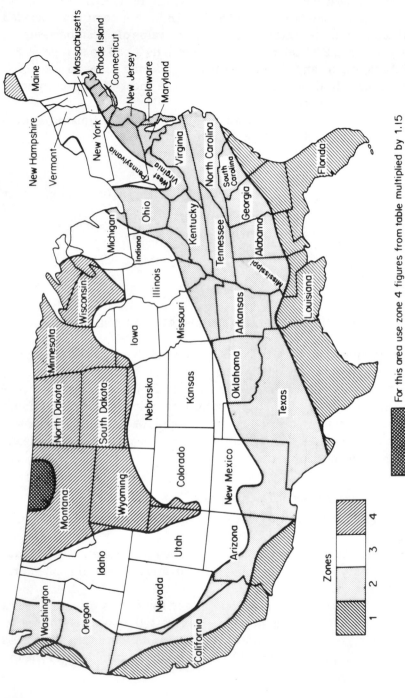

For this area use zone 4 figures from table multiplied by 1.15

Zones

1 2 3 4

FIGURE 7-12   Heating zones (*Courtesy* General Electric Company).

220

	Watts					Watts			
Square feet	Zone 1	Zone 2	Zone 3	Zone 4	Square feet	Zone 1	Zone 2	Zone 3	Zone 4
50	360	500	600	750	300	1650	2300	2750	3450
75	500	700	850	1050	325	1750	2450	2950	3675
100	650	900	1075	1350	350	1900	2650	3175	3975
125	750	1050	1275	1575	375	2000	2800	3350	4200
150	900	1250	1500	1875	400	2150	3000	3600	4500
175	1050	1450	1750	2175	450	2375	3300	3950	4950
200	1150	1600	1925	2400	500	2675	3700	4450	5550
225	1300	1800	2150	2700	550	2975	4000	4800	6000
250	1400	1950	2350	2925	600	3175	4400	5300	6600
275	1550	2150	2575	3225					

FIGURE 7-13   Approximate heat loss of various size rooms (*Courtesy* General Electric Company).

For rooms with more than one exterior wall (or wall exposed to unheated areas), the following multipliers must be applied:

1. *Two exterior walls:* multiply wattage obtained from Fig. 7-13 by 1.17.
2. *Three exterior walls:* multiply wattage obtained from Fig. 7-13 by 1.33.

7-6.2 Example of short-method heat-loss calculations

Estimate the heat loss of a room 14 ft long by 12 ft wide with two exterior (exposed) walls in Hastings, Nebraska.

1. Obtain the zone number for Hastings, Nebraska, from Fig. 7-12. Hastings, Nebraska, is in zone 3.
2. Find the square feet of floor area in the room. It is 14 $\times$ 12, or 168 ft^2.
3. From Fig. 7-13 obtain the wattage required to heat a room 169 ft^2 with one exposed wall in zone 3. Figure 7-13 does not include a room of this exact floor area. In this case it is advisable to use a room of the next larger size. In this example, use 175 ft^2. Under the zone 3 column, in line with a room of 175 ft^2, the heating requirement is shown as 1750 W.
4. Since the room has two exposed walls, the multiplier of 1.17 must be applied. The estimated heat loss is 1.17 $\times$ 1750, or 2048 W.
5. The total heating requirements of a home is obtained by

adding up the requirements of each room (from Fig. 7-13), with the necessary multiplier applied.

An alternative, even quicker, method for finding the total heating requirements for a complete home is given in Section 7-6.3.

7-6.3 Example of short-method heating-requirement calculations

In Fig. 7-14 the heating requirements are expressed in kilowatts for various sized homes of conventional, fully insulated frame-on-slab construction in areas of $0°$ design temperature. However, the requirements are comparable for similar construction, such as brick veneer or homes with basements.

The outdoor design temperature for a given city or town can be obtained from the local utility company, *ASHVE Guide*, and other heating publications. For example, the outdoor design temperature

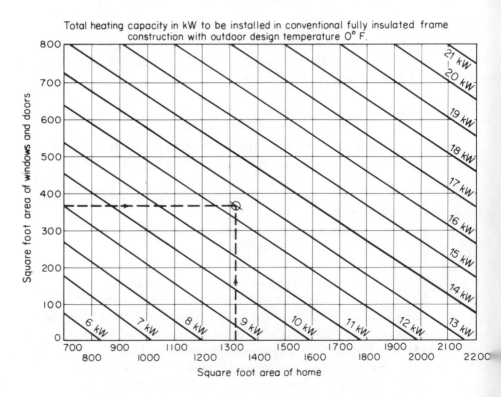

FIGURE 7-14   Total heating capacity table (*Courtesy* General Electric Company).

is 10° for Birmingham, Alabama; 0° for Washington, D.C.; and -10° for Chicago. It is important that the proper design temperature be used in determining the heating requirements.

Figure 7-14 should be used only to obtain the *approximate heating requirements of fully insulated homes.* For estimating kilowatt-hour consumption and operating cost, refer to Section 7-6.4.

The following information is required to use Fig. 7-14.

1. The total area of the home in square feet.
2. The total area of all windows and exterior doors in square feet.
3. The outdoor design temperature.

In areas of outdoor design temperatures above or below 0° the following correction factors must be applied:

Design Temp. (°F)	Multiplier
-10	1.16
0	1.00
10	0.86
20	0.72
30	0.57
40	0.43

Example

Estimate the heating requirements for a conventional fully insulated frame-on-slab home in a –5° design area. The home covers 1325 ft^2 and has 360 ft^2 of window and exterior door area.

1. On Fig. 7-14 draw a vertical line from the point representing 1325 ft^2 of floor area.
2. Draw a horizontal line from the point representing 360 ft^2 of window and door area.
3. Where the two lines intersect, read the heating requirements with reference to the diagonal lines. In this example the lines intersect at 12.4 kW.

To correct for a –5° design, multiply 12.8 × 1.08. The correction factor (1.08) is obtained by interpolating between the values at 0° and –10°.

The estimated heating requirements equal 12.4 × 1.08, or 13.39 kW.

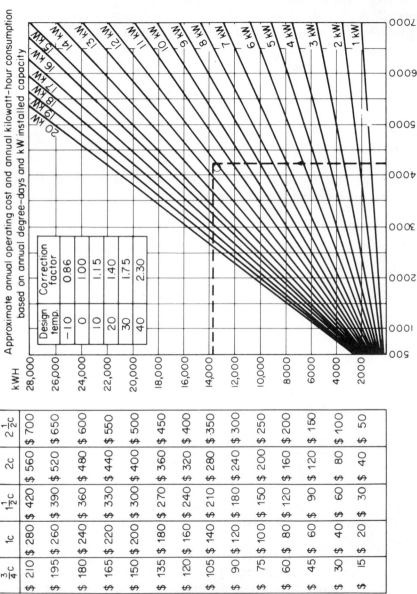

FIGURE 7-15 Heating cost table (*Courtesy* General Electric Company).

224

7-6.4 Short-form operating-cost calculations

The information of Fig. 7-15, when used with the procedures of Section 7-6.3, provides a means of quickly obtaining approximate annual kilowatthour consumption and operating cost information. Figure 7-15 can also be used with the procedures of Section 7-6.2.

Knowing the design temperature, degree-days, and required heating capacity in kilowatts, Fig. 7-15 can be used to obtain approximate values. In areas with outdoor design temperatures above or below 0°, the correction factors listed in Fig. 7-15 must be applied.

Example

Find the approximate annual kilowatthour consumption and operating cost for a home requiring 12.4 kW of heating capacity in an area of 0° design, with 4200 degree-days and a $1\frac{1}{2}$ cents/kWh electric rate.

From the 4200-degree-day point, draw a vertical line to the 12.4-kW point. Then draw a horizontal line to the left and read the approximate annual kilowatthour consumption and operating cost, 13,800 kWh and $207, as shown under the $1\frac{1}{2}$ cent rate column.

7-7 BRANCH CIRCUIT REQUIREMENTS FOR ELECTRIC HEATING

The branch circuit requirements for electric heating are *basically* the same as for any other load. Generally, the loads are pure resistive; thus the power factor is always 1.0. A possible exception is where the electric heater includes a fan. The fan motor may produce some reactive load (Chapter 9). The calculations for conductor size, voltage drop, and so on are as discussed in Chapters 1 through 5.

In addition to the basic wiring design, the following points should be considered when planning branch circuits for any electric heating system.

Of course, all wiring must be installed in accordance with the National Electric Code and local codes. If a nonmetallic system is used (nonmetal raceway), a third wire or other grounding means must be installed to provide adequate grounding for the electric heating equipment. Refer to Chapter 4.

Branch circuit wiring for lighting and other outlets, including branch circuits for the heating system, should *not* be routed adjacent to heated surfaces, or be subjected to temperatures higher than normal ambients. If it is necessary to install branch circuits so close that temperatures are higher than the normal ambients, conductors must be provided in accordance with the NEC or governing local code in the area.

With embedded wire heating units, if branch circuit wiring is enclosed in insulation in the ceiling or in a ceiling between floors in a multistory building, the branch circuit conductors must be supported at least 2 in. above the surface. Before installing embedded wire heating units, the contractor should check with the local inspector.

The NEC permits more than one heating unit to be supplied by a single branch circuit up to the allowable capacity of the branch circuit. Type T or TW (60°C) wire may be used for supply connections to all heating units, unless otherwise noted.

In most installations it is advisable to install a separate branch circuit for the heating system in each room, and to protect the branch circuit with a fuse rated the next size larger than the load (heater). When using circuit breakers it is advisable to combine rooms so as to approach the nominal capacity of the branch circuit and circuit breaker.

7-8 TEMPERATURE CONVERSIONS

Figure 7-16 provides for conversions of centigrade, Fahrenheit, Kelvin, and Rankine temperature units.

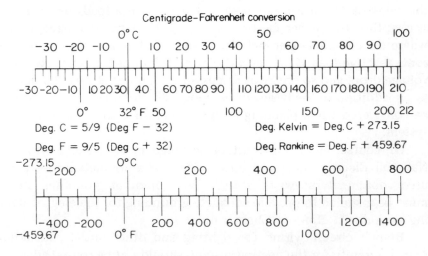

FIGURE **7-16   Simplified temperature conversion charts (***Courtesy*** Gen-eral Electric Company).**

Chapter 8

Electric Motor Wiring Design

From a mechanical design standpoint, an electric motor is considered to be a source of energy and is usually rated in such terms as horsepower, foot-pounds of torque, speed, etc. From an electrical design standpoint, an electric motor is simply another electrical load. That is, an electric motor requires a certain voltage and draws a certain amount of current. If there are several motors and they all require the same voltage, they can all be connected in the distribution system in parallel.

Electric motors usually have some inductive reactance. Therefore, the load presented by an electric motor is an impedance rather than a pure resistance, and a power factor is involved. Motors may be rated by voltage and current, or by voltage and wattage, with the power factor specified. Generally, motors have a lagging current since they are inductive. However, in certain circumstances, a synchronous motor will draw a leading current. This makes it possible to use synchronous motors to correct a lagging power factor condition or to bring a power factor within tolerance. This is discussed fully in Chapter 9.

Once the voltage and current for an electric motor (or group of motors) is established, the distribution system can be designed to provide the necessary power, using the information of Chapters 1 through 4 (conductor size, voltage drop, grounding, overcurrent protection, etc.) If different voltages are required for different motors in the system, transformers can be used, as discussed in Chapter 5.

In addition to designing the power distribution system to the motors, it is necessary to provide the motors with *control devices.* Even the simplest motors must have a means of starting and stopping, as well as overcurrent protection (in addition to that provided by the branch circuits of the distribution system). It is generally not necessary to design the motor control components (starters, controllers etc.), because these are available as off-the-shelf items (as are the motors). However, it is necessary to understand the operation of the motor control systems to provide the correct wiring and to select the correct motor control components.

The purpose of this chapter is to familiarize the electrician or contractor with motor control fundamentals. The basic theory of electric motors, starters, controllers, and relays is omitted. Instead, we shall concentrate on those electric motor characteristics that directly and indirectly affect the design of electrical wiring. For example, a certain class of starter is required for a motor given the voltage and current. Such data are available in chart or tabular form. This simplifies the designer's task when specifying motor starters.

This chapter also provides definitions, symbols, diagrams, and illustrations that give the reader a sound background in the language and basic principles associated with motor control components. The material in this chapter is limited to ac motors, operating at voltages less than 600 V. Direct-current and high-voltage motors have very special design requirements.

8-1 ELECTRIC MOTOR BASICS

Before going into the practical design of electric motor wiring, let us review the basics of electric motors—the terms used, the most common types, and characteristics.

8-1.1 The squirrel-cage motor

The great majority of ac motors in use today are of the squirrel-cage type. This motor gets its name from the rotor construction. The rotor has no wire windings. Rotor bars are used instead, and the rotor has ball bearings rather than sleeve bearings. As the rotor bars are cut by the stator flux, the bars have a voltage induced by transformer action. A current will flow in the short-circuited rotor bars, causing a magnetic flux around the bars. This develops a torque, causing the rotor to follow the rotating field. Squirrel-cage motors are simple in construction and operation. Simply connect three power lines to the motor, and it will run.

8-1.2 Motor speed and slip

The speed of a squirrel-cage motor depends on the number of poles in the motor's winding. With 60-Hz power, a two-pole motor runs at about 3450 r/min, a four-pole at 1725 r/min, and a six-pole at 1150 r/min. Motor nameplates are usually marked with these actual *full load speeds*. However, motors are frequently referred to by their *synchronous speeds:* 3600, 1800, 1200 r/min, respectively.

The difference between synchronous speeds and full load speeds can be explained as follows. The starting current of a squirrel-cage motor is high but of very short duration and decreases as the rotor approaches the speed of the rotating field. The rotor cannot reach this synchronous speed. If it could, no flux would cut the rotor, there would be no induction or rotor current or torque, and the motor would slow down. The actual rotor speed slips behind the rotating field sufficiently so it can induce enough current to produce the torque needed to satisfy the demands of the mechanical load.

The inability to keep up with the synchronous speed is an important measure of a motor's performance and is called *slip*. Slip may be measured in r/min or as a percentage of synchronous speed. When written as a decimal, the symbol for slip is *s*.

8-1.3 Motor torque

Torque is the "turning" or "twisting" force of the motor and is usually measured in pound-feet (or foot-pounds). Except when the motor is accelerating up to speed, the torque is related to the motor horsepower by the equation

$$\text{torque in pound-feet} = \frac{\text{hp} \times 5252}{\text{r/min}}$$

The torque of a 25-hp motor running at 1725 r/min would be computed as follows:

$$\text{torque} = \frac{25 \times 5252}{1725} = \text{approx 76 lb-ft}$$

If 90 lb-ft was required to drive a particular load, the motor would be overloaded and would draw a current in excess of the *full load current* (FLC), the current required to produce full load torque at the rated speed.

8-1.4 Motor temperature ratings

Both *ambient temperature* and *temperature rise* must be considered when planning electric motor systems.

The *ambient temperature* is the temperature of the air where the motor or its control equipment is operating. Most motor controllers are of the enclosed type, and the ambient temperature is the temperature of the air *outside the enclosure*, not inside. Likewise, if a motor has an ambient temperature of 30°C (86°F), this is the temperature of the air outside the motor. Except in very special applications, motors and controllers manufactured to NEMA standards are subject to a 40°C (104°F) ambient temperature limit.

No matter what ambient temperature exists when the motor starts, the temperature will rise after the motor is running. This temperature increase is because of the current passing through the motor windings. The difference between the winding temperature of the motor (when running) and the ambient temperature is called the *temperature difference*.

The temperature rise produced at full load is not harmful provided that the motor ambient temperature does not exceed 40°C (104°F). Higher temperature caused by increased current, or higher ambient temperatures, produces a deteriorating effect on motor insulation and lubrication.

A rule of thumb states that for each increase of 10° above the rated temperature, motor life is cut in half.

8-1.5 Motor service factor

If the motor manufacturer has given a motor a *service factor*, it means that the motor can be allowed to develop more than its rated or nameplate horsepower without causing undue deterioration of the insulation. The service factor is a margin of safety. For example, if a 10-hp motor has a service factor of 1.15, the motor *can be allowed* to develop 11.5 hp, although there is no assurance that the motor will develop more than 10 hp. The service factor is a matter of motor design.

8-1.6 Motor duty ratings (time ratings)

Most motors have a *continuous duty rating.* This permits indefinite operation at a rated load.

Intermittent duty ratings are based on a fixed operating time (5, 15, 30, 60 min) after which the motor must be allowed to cool (usually for a specific period of time).

Some motors have both continuous and intermittent duty ratings. That is, they will operate continuously at a given load and intermittently at higher loads.

8-1.7 Plugging, jogging, and inching

The term *plugging* is used when a motor running in one direction is momentarily reconnected to reverse the direction and is brought to rest very rapidly. *Jogging* (also known as *inching*) describes the repeated starting and stopping (but not reversing) of a motor at frequent intervals for short periods of time. If either plugging or jogging is to occur more frequently than 5 times per minute, the starter or controller must be *derated.*

For example, a NEMA size 1 starter has a normal duty rating of $7\frac{1}{2}$ hp at 230 V (polyphase). With jogging or plugging, the same starter has a maximum rating of 3 hp. From a design standpoint, if the motor must deliver the full $7\frac{1}{2}$ hp with jogging or plugging, a larger starter is required. In this case a NEMA size 2 starter would be required. The subject of starter and controller ratings from a design standpoint is discussed in later sections of this chapter.

8-1.8 Sequence (interlocked) control

Many processes require a number of separate motors that must be started and stopped in a definite sequence, such as a system of conveyors. When starting up the delivery conveyor must start first with the other conveyers starting in sequence, to avoid pileup of material. When shutting down, the reverse sequence must be followed with time delays between the shutdowns (except for emergency stops) so that no material is left on the conveyors. This is an example of simple *sequence control* (sometimes known as *interlocked control*).

Separate starters could be used, but it is common to build a special controller that incorporates starters for each motor, as well as timers, control relays, and so on, to accomplish the timing sequence. From a design standpoint, such controllers are generally "special order equipment"; that is, they are built to the designer's specifications (of timing, sequence, etc.) by the manufacturer.

8-1.9 Locked rotor current

During the acceleration period at the moment a motor is started, it draws a high current, called the *inrush current*. The inrush current when the motor is connected directly to the line (so that the full line voltage is applied to the motor) is called the *locked rotor current* (LRC) or the stalled rotor current (SRC). The LRC can be from about 4 to 10 times the motor full load current (FLC, Section 8-1.3). Most motors have an LRC of about 6 times FLC, and therefore this figure is generally used. The 6-times value is often expressed as 600 per cent of FLC.

8-1.10 Motor controllers, starters, switches, contactors, and relays

All the equipment listing in this heading can be considered to be *motor control equipment*. The terms are often used interchangeably. Equally often, an improper term is used. For example, contactors and magnetic controllers are often confused. For that reason, we shall establish the following definitions for the terms used in this chapter. Each of the devices are discussed in separate paragraphs.

A motor *switch* provides only an on–off function for the motor. However, some motor switches also provide for reversing direction of the motor.

A *motor controller* will include some or all of the following functions: starting, stopping, overload protection, overcurrent protection, reversing, changing speed, jogging, plugging, sequence control, and possibly pilot-light indication. A motor controller can also provide control of auxiliary equipment, such as breaks, clutches, solenoids, heaters, and signals. A motor controller may be used to control a single motor or a group of motors.

The terms *starter* and *controller* mean practically the same thing. Strictly speaking, a starter is the simplest form of controller and is capable of starting and stopping the motor and providing it with overload protection. The starter shown in Fig. 8-1 could qualify as a controller since it provides on–off (by means of a switch) and overload protection (by means of a thermal overload element to the left of the switch).

Motor controllers and starters can be either *manual* or *magnetic*.

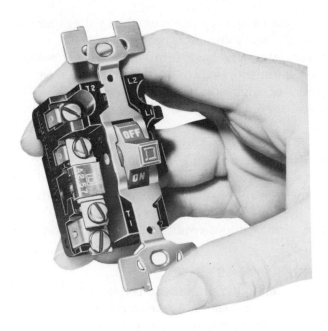

FIGURE 8-1   Starter switch includes overcurrent protection (*Courtesy* Square D Company).

The adjective "manual" usually applies only to starters. However, both starters and controllers can be magnetic.

The general classification *contactor* covers a type of magnetically operated device designed to handle relatively *high currents*. A special form of contactor exists for lighting equipment. Lighting contactors are discussed in Section 8-7.2.

The conventional motor contactor is identical in appearance, construction, and current-carrying ability to the equivalent NEMA size magnetic starter. The magnet assembly and coil, contacts, holding circuit interlock, and other structural features are the same.

The significant difference is that the contactor *does not provide overload protection*. Contactors are thus used to switch high-current nonmotor loads or are used in motor circuits if overload protection is separately provided.

A control *relay* is also a magnetically operated device, similar in operating characteristics to a contactor. However, the relay is used to switch low-current circuits. Relays are used in *control circuits*, where little current is needed (typically 15 A at 600 V), whereas contactors are used in the *power circuit* (line voltage directly to motor windings), where heavy current is needed.

Contactors generally have from one to five poles. Although normally open and normally closed contacts can be provided, the great majority of applications use the normally open contact configuration and there is little, if any, conversion of contact operation in the field.

As compared to contactors, it is not uncommon to find relays used in applications requiring 10 to 12 poles per device, with various combinations of normally open and normally closed contacts. In addition, some relays have convertible contacts, permitting changes to be made in the field from normally open to normally closed operation, or vice versa, without requiring kits or additional components.

8-1.11 Motor controller enclosures

The NEMA and other organizations have established standards of enclosure construction for motor control equipment. In general, motor control equipment is enclosed for one or more of the following reasons:

1. Prevent accidental contact with live parts.
2. Protect the control from harmful environmental conditions.
3. Prevent explosion or fires that might result from the electrical arc caused by the control.

Common types of enclosures per NEMA classification numbers are:

NEMA 1 — general purpose. The general-purpose enclosure, shown in Fig. 8-2, is intended to prevent accidental contact with the enclosed apparatus. It is suitable for general-purpose applications indoors, where it is not exposed to unusual service conditions. A NEMA 1 enclosure serves as protection against dust and light indirect splashing but is not dusttight.

Generally, when the type of enclosure is not specified, the designer can use a NEMA 1. If specific operating conditions are given, the designer must use the corresponding type of enclosure. The following is a summary of the common NEMA types. The NEMA standard literature must be consulted for a full description of the types. However, the titles for each type are generally self-explanatory.

NEMA 3 — dusttight, raintight.

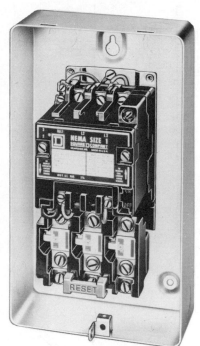

FIGURE 8-2   NEMA 1 General Purpose Enclosure (*Courtesy* Square D Company).

NEMA 3R — rainproof, sleet resistant.

NEMA 4 — watertight.

NEMA 4X — watertight, corrosion resistant.

NEMA 7 — hazardous locations, Class I (meets NEC Class I hazardous location standards).

NEMA 9 — hazardous locations, Class II (meets NEC Class II hazardous location standards).

NEMA 12 — industrial use.

NEMA 13 — oiltight, dusttight.

8-2 WIRING DIAGRAMS USED IN MOTOR CONTROL CIRCUITS

Before going into the details of motor control characteristics, we shall discuss the basics of wiring diagrams used in motor control circuits.   Both *wiring diagrams* and *elementary* or *schematic diagrams* are used.

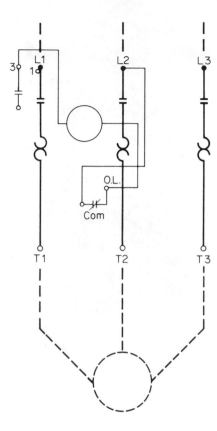

FIGURE **8-3**   Typical wiring diagram (*Courtesy* Square D Company).

Wiring diagrams, such as shown in Fig. 8-3, show, as closely as possible, the actual location of all the component parts of the system. In the case of the circuit in Fig. 8-3, the system shown is a motor starter. The dotted lines represent power circuit connections made to the starter, and from the starter to the motor.

Since wiring connections and terminal markings are shown, the wiring diagram is helpful when wiring the starter or when tracing wires when troubleshooting. Note that bold lines denote the power circuit, and thin lines are used to show the control circuit. Conventionally, in ac magnetic equipment, black wires are used in power circuits and red wiring is used for control circuits.

A wiring diagram is limited in its ability to convey a clear picture of the sequence of operation of a controller, starter, and so on. Where an illustration of the circuit in its simplest form is desired, the elementary diagram or schematic diagram is used.

The *elementary*, or *schematic diagram*, such as shown in Fig. 8-4,

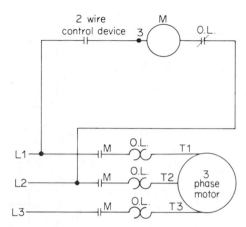

FIGURE 8-4   Typical elementary (schematic) diagram   (*Courtesy* Square D Company).

gives a fast, easily understood picture of the circuit. The devices and components are not shown in their actual positions. All the control circuit components are shown as directly as possible, between a pair of vertical lines representing the control power supply. The arrangement of the components is designed to show the sequence of operation of the devices and helps in understanding how the circuit operates. The effect of operating various control devices can be readily seen. This helps in troubleshooting, particularly with the more complex controllers.

A summary of the symbols used in wiring diagrams and schematic diagrams is given in Fig. 8-5.

SUPPLEMENTARY   CONTACT   SYMBOLS

SPST N.O.		SPST N.C.		SPDT		TERMS
SINGLE BREAK	DOUBLE BREAK	SINGLE BREAK	DOUBLE BREAK	SINGLE BREAK	DOUBLE BREAK	SPST – SINGLE POLE SINGLE THROW
						SPDT – SINGLE POLE DOUBLE THROW
DPST. 2 N.O.		DPST. 2 N.C.		DPDT		DPST – DOUBLE POLE SINGLE THROW
SINGLE BREAK	DOUBLE BREAK	SINGLE BREAK	DOUBLE BREAK	SINGLE BREAK	DOUBLE BREAK	DPDT – DOUBLE POLE DOUBLE THROW
						N.O. – NORMALLY OPEN
						N.C. – NORMALLY CLOSED

FIGURE 8-5   Standard Elementary Diagram Symbols (*Courtesy* Square D Company).

The diagram symbols shown below have been adopted by the Square D Company and conform where applicable to standards established by the National Electrical Manufacturers Association (NEMA).

SWITCHES

DISCONNECT	CIRCUIT INTERRUPTER	CIRCUIT BREAKER W/THERMAL O.L.	CIRCUIT BREAKER W/MAGNETIC O.L.	CIRCUIT BREAKER W/THERMAL AND MAGNETIC O.L.	LIMIT SWITCHES		FOOT SWITCHES	
					NORMALLY OPEN	NORMALLY CLOSED	N.O	N.C

HELD CLOSED   HELD OPEN

PRESSURE & VACUUM SWITCHES		LIQUID LEVEL SWITCH		TEMPERATURE ACTUATED SWITCH		FLOW SWITCH (AIR, WATER, ETC.)	
N.O.	N.C.	N.O.	N.C.	N.O.	N.C.	N.O.	N.C.

SPEED (PLUGGING)

ANTI-PLUG

F   R

F   R

F   R

SELECTOR

2 POSITION

J   K

	J	K
A1	1	
A2		1

1—CONTACT CLOSED

3 POSITION

J   K   L

	J	K	L
A1	1		
A2		1	1

1—CONTACT CLOSED

2 POS. SEL. PUSH BUTTON

A   B

	SELECTOR POSITION			
	A		B	
CONTACTS	BUTTON		BUTTON	
	FREE	DEPRES'D	FREE	DEPRES'D
1—2	1			1
3—4		1	1	

1—CONTACT CLOSED

FIGURE 8-5   (Continued)

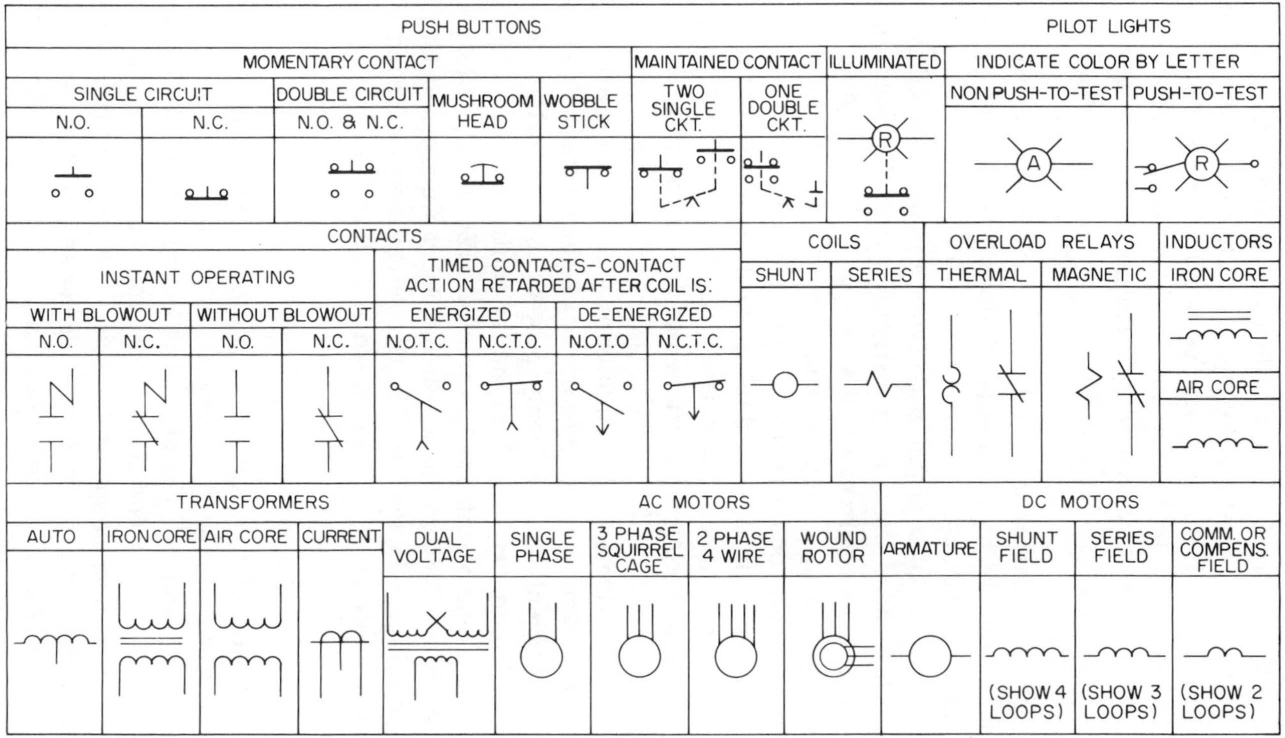

FIGURE 8-5 **(Continued)**

8-3 PROTECTING ELECTRIC MOTORS

Motors can be damaged, or their effective life reduced, when subjected to continuous current only slightly higher than their full load current (Section 8-1.3), times the service factor (Section 8-1.5). Motors are designed to handle inrush or locked rotor currents (Section 8-1.9) without excessive temperature rise, provided the acceleration time is not too long or the duty cycle (see jogging, Section 8-1.7) too frequent.

Damage to insulation and windings of the motor can also be sustained on extremely high currents of short duration, as found in grounds or short circuits.

All currents in excess of full load current can be classified as *overcurrents*. In general, however a distinction is made based on the magnitude of the overcurrent and the equipment to be protected.

An overcurrent up to locked rotor current is usually the result of *mechanical overload* on the motor. The subject of protection against this type of overcurrent is covered in Article 430 (Part C) of the NEC, entitled "Motor Running Overcurrent (Overload) Protection." In this chapter the designation is shortened to "overload protection" and is covered in Section 8-3.2.

Overcurrents due to short circuits or grounds are much higher than the LRC. Equipment used to protect against damage due to this type of overcurrent must not only protect the motor but also the branch circuit conductors and the motor controller. Provisions for the protective equipment are specified in NEC Article 430 under Part D, entitled "Motor Branch Circuit Short Circuit and Ground Fault Protection." In this chapter the designation is shortened to "overcurrent protection" and is covered in Section 8-3.1.

8-3.1 Overcurrent protection

The function of the overcurrent protective device is to protect the motor branch circuit conductors, control apparatus, and motor from shorts or grounds. The protective devices commonly used are thermal-type magnetic circuit breakers and fuses. *From a design standpoint, the short-circuit device must be capable of carrying the starting current of the motor, but the device setting must not exceed 250 per cent of full load current (with no code letter on the motor), or from 150 to 250 per cent of full load current, depending upon the code letter of the motor.* Where the value is not sufficient to carry

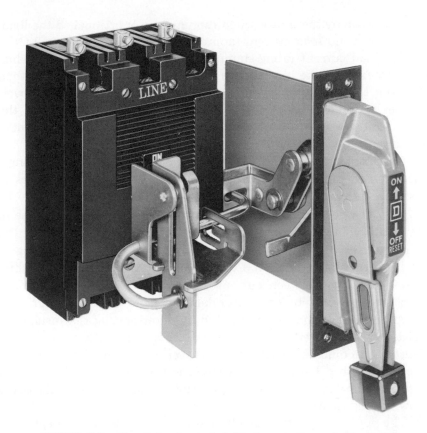

FIGURE 8-6   Circuit breaker includes fault protection and disconnect required by NEC (*Courtesy* Square D Company).

the starting current, it may be increased *but must in no case exceed 400 per cent* of the motor full load current.

The NEC requires (with few exceptions) a means to disconnect the motor and controller from the line, in addition to an overcurrent protective device to clear short-circuit faults. The circuit breaker shown in Fig. 8-6 incorporates fault protection (circuit breakers) and disconnect (lever switch) in one basic device. When the overcurrent protection is provided by fuses, a disconnect switch is required and the switch and fuses are generally combined in one housing.

8-3.2 Overloads

A motor has no intelligence and will attempt to drive any load, even if excessive. Exclusive of inrush or LRC when accelerating, the current drawn by the motor when running is proportional to the load, varying from *no load* current (approximately 40 per cent of FLC) to the full load current rating stamped on the motor nameplate. When the load exceeds the torque rating of the motor, it draws higher than FLC and the condition is described as overload. The maximum overload exists under LRC, in which the load is so excessive that the motor stalls or fails to start and, as a consequence, draws continual inrush current or LRC.

Overloads can be electrical as well as mechanical in origin. Single phasing of a polyphase motor and low line voltage are examples of electrical overload.

8-3.2.1 Overload protection. The effect of an overload is a rise in temperature in the motor windings. The larger the overload, the more quickly the temperature will increase, to a point damaging to the insulation and lubrication of the motor. An inverse relationship, therefore, exists between current and time; the higher the current, the shorter the time before motor damage or burnout can occur.

All overloads shorten motor life by deteriorating the insulation. Relatively small overloads of short duration cause little damage, but, if sustained, could be just as harmful as overloads of greater magnitude. The relationship between the magnitude (per cent of full load) and duration (time in minutes) of an overload is illustrated by the curve shown in Figure 8-7.

The ideal overload protection for a motor is an element with current-sensing properties very similar to the heating curve of the motor, which would act to open the motor current when FLC is exceeded. The operation of the protective device should be such that the motor is allowed to carry harmless overloads but is quickly removed from the line when an overload has persisted too long.

8-3.2.2 Overload protection with fuses. Fuses are not designed to provide overload protection. Their basic function is to protect against short circuits (overcurrents, Section 8-3.1). Motors draw a high inrush current (generally 6 times the normal FLC) when starting. Single-element fuses have no way of distinguishing between this temporary and harmless inrush current and a damaging overload. Thus a fuse chosen on the basis of motor FLC would blow every time the motor started. On the other hand, if a fuse large enough to pass the starting or inrush current is chosen, it would not protect the motor against small, but harmful, overloads that might occur later.

Dual-element or time-delay fuses can provide motor-overload pro-

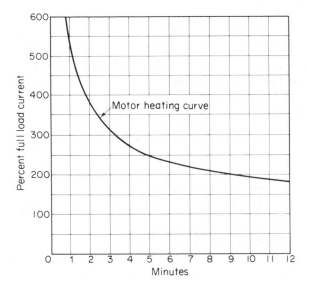

Application of Motor Heating Curve Data
On 300% OVERLOAD, the particular motor for which this curve is characteristic would reach its permissable temperature limit in 3 minutes. Overheating or motor damage would occur if the OVER- LOAD persisted beyond that time.

FIGURE 8-7   Typical motor heating curve (*Courtesy* Square D Company).

tection but suffer the disadvantage of being nonrenewable and must be replaced.

8-3.2.3   **Overload protection with relays.**   The overload relay is the heart of motor protection. Like the dual-element fuse, the overload relay has inverse-trip-time characteristics, permitting it to hold in during the accelerating period (when inrush current is drawn), yet providing protection on a small overload above FLC when the motor is running. Unlike the fuse, the overload relay is *renewable* and can withstand repeated trip and reset cycles without need of replacement. It should be emphasized that the overload relay does not provide short-circuit protection, as this is the function of the overcurrent protective equipment such as fuses and circuit breakers (Section 8-3.1) at the branch circuits.

The overload relay consist of a current-sensing unit connected in the line to the motor, plus a mechanism, actuated by the sensing unit, which serves to directly or indirectly break the circuit. In a manual starter (Section 8-4), an overload trips a mechanical latch, causing the starter contacts to open and disconnect the motor from the line. In magnetic starters (Section 8-5), an overload opens a set of contacts within the overload relay itself. These contacts are wired in series with the starter coil in the control circuit of the magnetic starter. Breaking the coil circuit causes the starter contacts to open, disconnecting the motor from the line.

Overload relays can be classified as *thermal* (Section 8-3.2.4) or *magnetic* (Section 8-3.2.5). Magnetic overload relays react only to current excesses and are not affected by temperature. As the name implies, thermal overload relays rely on the rising temperatures caused by the overload current to trip the overload mechanism. Thermal overload relays can be further subdivided into two types, melting alloy and bimetallic.

8-3.2.4 Thermal overload relays.

Melting-alloy thermal overload relays. In a melting-alloy thermal overload relay (also referred to as a *solder pot relay*) the motor current passes through a small heater winding. Under overload conditions, the heat causes a special solder to melt, allowing a ratchet wheel to spin free, opening the contacts. When this occurs the relay is said to trip.

To obtain appropriate tripping current for motors of different sizes, or differential full load currents, a range of thermal units (heaters) is available. The heater coil and solder pot are combined in a one-piece nontamperable unit. The heat-transfer characteristics and the accuracy or the unit cannot be accidentally changed as is possible when the heater is a separate component. Melting-alloy thermal overload relays have to be reset by hand; thus after they trip, they must be reset by a deliberate hand operation. A reset button is usually mounted on the cover of enclosed starters. *From a design standpoint, thermal units are rated in amperes and are selected on the basis of motor FLC, not on horsepower.*

Bimetallic thermal overload relays. A bimetallic thermal overload relay uses a U-shaped bimetal strip associated with a current-carrying heater element. When an overload occurs, the heat will cause the bimetal to deflect and open a contact. Different heaters give different trip points. In addition, most relays are adjustable over a range of 85 to 116 per cent of the nominal heater rating.

Bimetal relays are field convertible from *hand reset* to *automatic reset*, and vice versa. On automatic reset, the relay contacts, after tripping, will automatically reclose when the relay has cooled off. This is an advantage when the relays are inaccessible.

However, automatic reset overload relays should not normally be used with two-wire control. With this arrangement, when the overload relay contacts reclose after an overload relay trip, the motor will restart and, unless the cause of the overload has been removed, the overload relay will trip again. This cycle will repeat, and eventually the motor will burn out, as a result of the accumulated heat from the repeated inrush current. More important is the possibility of danger to personnel. The unexpected restarting of a machine when the operator or maintenance person is attempting to find out why the machine has stopped places the person in a hazardous situation.

Ambient-compensated bimetallic overload relays. Ambient-compensated bimetallic overload relays are designed for one single situation, when the motor is at a constant temperature and the controller is *located separately at a varying temperature.* In this case, if a standard thermal overload relay is used, the relay will not trip consistently at the same level of motor current if the controller temperature changes. To compensate for the temperature variation the controller may see, an ambient-compensated overload relay is applied. Its trip point is *not affected* by temperature and it performs consistently at the same value of current.

Both *melting-alloy* and *bimetallic* overload relays are designed to approximate the heat actually generated in the motor. As the motor temperature increases, so does the temperature of the thermal unit in the relay. The motor and relay heating curves (Fig. 8-8) show this relationship. From the graph we can see that no matter how high the current drawn, the overload relay will provide protection, yet the relay will not trip unnecessarily.

In those cases where it is necessary to signal an overload condition, the overload relay is fitted with a set of contacts that close when the relay trips, thus completing an alarm circuit. These contacts are called *alarm contacts.*

Selecting thermal overload relays. From a design standpoint, motor FLC, the type of motor, and the possible difference in ambient temperature must all be taken into account when choosing overload relay thermal units or overload heaters. Motors of the same horsepower and speed do not all have the same full load current. *Always refer* to the motor nameplate for the FLC; do not use a published

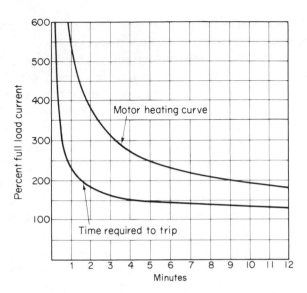

Graph shows motor heating curve and overload relay trip curve.
Overload relay will always trip out at a safe value.

FIGURE 8-8   **Motor heating curve versus trip characteristics of overload (*Courtesy* Square D Company).**

table. These tables of motor FLC show the *average* or *normal* FLC.
The FLC of the motor in question may be quite different, particularly
with the small single-phase motors.

Relay thermal unit selection tables are published on the basis of
continuous-duty motors, with a 1.15 service factor, operating under
normal conditions. The tables shown in catalogs also appear (usually)
on the inside of the floor or cover of the controller. The selections
for relay thermal units shown in the table will properly protect the
motor and permit the motor to develop its full horsepower, allowing
for a service factor, *if the ambient temperature is the same at the
motor as at the controller.* If not, or if the service factor is not the
standard 1.15, consult the motor/controller manufacturer for data or
special procedures to select a proper thermal unit.

8-3.2.5 Magnetic overload relays. A magnetic overload relay such as
the one shown in Fig. 8-9 has a movable magnetic core inside a coil
that carries the motor current. The flux set up inside the coil pulls
the core upward. When the core rises far enough (determined by the
current and the position of the core) it trips a set of contacts on the
top of the relay.

FIGURE 8-9   Magnetic overload relay
(*Courtesy* Square D Company).

The movement of the core is slowed by a piston working in an oil-filled dash pot (similar to a shock absorber) mounted below the coil. This produces an inverse-time characteristic. The effective tripping current is adjusted by moving the core on a threaded rod. The tripping time is varied by uncovering oil bypass holes in the piston. Because of time and current adjustments, the magnetic overload relay is sometimes used to protect motors that have long accelerating times or unusual duty cycles. (The instantaneous trip magnetic overload relay is similar but has no oil-filed dashpot.)

From a design standpoint, the main concern of any overload relay is that the coil will operate at the available voltage and the contacts will carry the current. Usually the contacts are rated for both alternating current and direct current. Similarly, the contacts are rated by continuous make and break currents. For example, a relay may have a continuous current rating of 10 A, a make rating of 7500 VA, and a break rating of 750 VA. If the voltage through the contacts is 240 V, the instantaneous current at the time contact is made is about 30 A (7500 VA/240 V), whereas the instantaneous break current is about 3 A (750 VA/240 V). With the motor operating continuously, the contacts can carry 10 A.

8-4 MANUAL STARTERS FOR ELECTRIC MOTORS

A manual starter is a motor controller whose contact mechanism is operated by a mechanical linkage from a toggle handle or push button, which is, in turn, operated by hand. A thermal unit and direct-acting overload mechanism provides motor running overload protection. Basically, a manual starter is an on–off switch with overload relays. The small fractional horsepower unit shown in Fig. 8-1 is a manual starter.

Manual starters are generally used on small machine tools, fans and blower, pumps, compressors, and conveyors. They are the lowest in cost of all motor starters, have a simple mechanism, and provide quiet operation with no "magnet hum." Moving a handle or pushing the "start" button closes the contacts, which remain closed until the handle is moved to "off," or the "stop" button is pushed, or the overload relay thermal unit trips.

Manual starters cannot provide low-voltage protection or low-voltage release as can magnetic controllers (Section 8-5). If the power fails, the contacts *remain closed*, and the motor will restart when the power returns. This is an advantage for pumps, fans, compressors, oil burners, and the like, but for other applications it can be a disadvantage and can even be dangerous to personnel or equipment.

From a design standpoint, a manual starter should not be used in an application of any type where the machine or operator will be endangered if power fails and then returns without warning. For such applications a magnetic starter and momentary-contact pilot device such as described in Section 8-5 should be used for safety purposes.

Manual starters are of the *fractional-horsepower* type, or the *integral-horsepower* type. Both types provide *across-the-line* starting. That is, the switches and overload element are in series with the power line (or conductors) to the motor.

8-4.1 Fractional-horsepower manual starter

Fractional-horsepower (FHP) *manual starters* are designed to control and provide overload protection for motors of 1 hp or less, on 120- or 240-V single-phase power. FHP starters are available in single-power and two-pole versions and are operated by a toggle handle on the front. Typical wiring diagrams are shown in Fig. 8-10.

When a serious overload occurs, the thermal unit trips to open the starter contacts, disconnecting the motor from the line. In this sense FHP manual starters are similar to circuit breakers. The contacts cannot be reclosed until the overload device has been reset by moving the handle to the "full off" position, after allowing about 2 min for

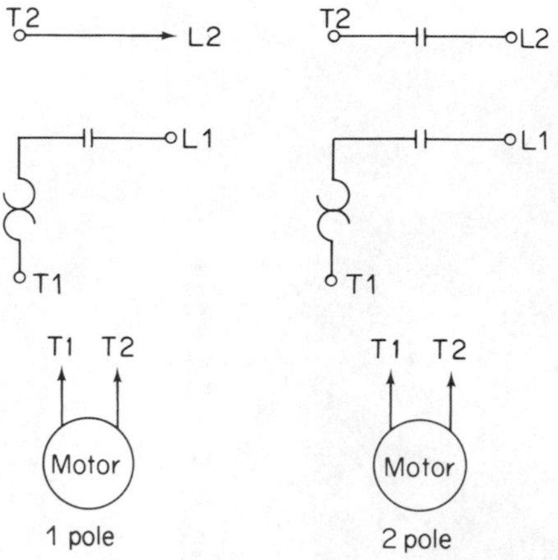

FIGURE 8-10   Typical manual starter wiring diagrams (*Courtesy* Square D Company).

the thermal unit to cool. The open type of manual starter (Fig. 8-1) will fit into a standard outlet box and can be used with a standard flush plate.

8-4.2 Integral-horsepower manual starter

The *integral-horsepower manual starter* is available in two- and three-pole versions, to control single-phase motors up to 5 hp and polyphase motors up to 10 hp. A typical integral-horsepower manual starter is shown in Fig. 8-11.

Two-pole starters have one overload relay; three-pole starters usually have two overload relays but are available with three overload relays. When an overload relay trips the starter mechanism unlatches, opening the contacts to stop the motor. The contacts cannot be reclosed until the starter mechanism has been reset by pressing the "stop" button or moving the handle to the "reset" position, after allowing time for the thermal unit to cool.

8-4.3 Manual motor starting switches

Manual motor starting switches provide on–off control of single-phase or three-phase motors where overload protection is *not required* or is provided separately. Typically, motor starter switches are used

FIGURE 8-11   Typical integral horsepower manual starter (*Courtesy* Square D Company).

where the branch circuits are adequately protected for overcurrents, as described in Chapter 3.

Two- or three-pole switches are usually available with ratings up to 5 hp 600 V three-phase. The continuous current rating is typically 30 A at 250 V maximum, and 20 A at 600 V maximum.

The toggle operation of the manual switch is similar to the FHP starter, and typical applications include small machine tools, pumps, fans, conveyors, and other electrical machinery that has separate motor protection.

Manual motor starting switches are often used with resistance heating units and other nonmotor loads that require the corresponding voltage and current ratings for the contacts.

8-5 MAGNETIC CONTROL FOR ELECTRIC MOTORS

A high percentage of applications require the controller to be capable of operation from remote locations or to provide automatic operation in response to signals from such pilot devices as thermostats, pressure or float switches, and limit switches. Thus magnetic control is used. A typical magnetic starter, NEMA size 1, is shown in Fig. 8-12.

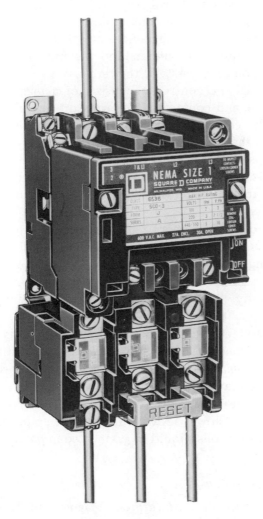

FIGURE 8-12   Typical magnetic starter (*Courtesy* Square D Company).

With manual control the starter must be mounted so that it is easily accessible to the operator. With magnetic control the push-button stations or other pilot devices can be mounted anywhere on the machine and connected by control wiring into the coil circuit of the remotely mounted starter.

8-5.1 Magnetic controller construction

Since we are concerned primarily with design, we shall not discuss construction of magnetic control devices in detail. Instead, we shall concentrate on those characteristics that most affect design. However, it is necessary to understand the basis of magnetic controller construction to understand the terms.

A typical magnetic controller consists of a magnet assembly, a coil, an armature, and contacts. The armature is controlled by current through the coil. The contacts are mechanically connected to the armature so that when the armature moves to its closed position, the contacts close. This applies power to the motor. When the coil has been energized and the armature and contacts have moved to the closed position, the controller is said to be *picked up* and the armature is *seated* or *sealed in.*

When the coil is first energized, it draws a fairly high current, known as the *inrush current.* As the armature and contacts move closer to closing, the coil current is reduced. When the contacts finally close, the current through the coil is known as the *sealed current.* Typically, the inrush current is 6 to 10 times the sealed current.

When the coil is first energized, the impedance is low. After seal in, the impedance is high. For this reason, ac magnetic coils should *never be connected in series.* If one controller seals in ahead of the other (which can occur even if they are identical, and will occur if they are not), the increased circuit impedance will reduce the coil current so that the "slow" device will not pick up or, having picked up, will not seal. Ac coils should be connected in *parallel.*

8-5.2 Magnetic controller ratings

Inrush and sealed current ratings. Magnet coil data are usually given in units of volt-amperes (volts times amperes, VA). For example, given a magnetic starter whose coils are rated at 600 VA inrush and 60 VA sealed, the inrush current of a 120-V coil is 600/120, or 5 A.

The same starter with a 480-V coil will only draw 600/480, or 1.25 A, inrush, and 60/480, or 0.125 A, sealed.

Magnet coil information is found in tables such the one shown in Fig. 8-13. This table is for magnet coils used in starters and controllers manufactured by the Square D Company and carries their

Standard AC Magnet Coils

Size	Type	Poles	Coil specification Number	Hz	6 Volts	12 Volts	24 Volts	110-115 Volts	120 Volts	208 Volts	220 Volts	240 Volts	277 Volts	440 Volts	480 Volts	550 Volts	600 Volts	Inrush	Sealed
— SUFFIX NUMBER (Complete Part Number of Coil Consists of Specification Number Followed by Suffix Number as 31041-400-42)																		Coil Volt-Amperes	

CLASSES 8501, 8502, 8536, 8538, 8539, 8547, 8549, 8606, 8630, 8640, 8650, 8651, 8702, 8736, 8738, 8739, 8810, 8811, 8812, 8903

Size	Type	Poles	Coil spec. No.	Hz	6	12	24	110-115	120	208	220	240	277	440	480	550	600	Inrush	Sealed
0	BH & BR B	2-4	1861-S1	60	R16A	R19B	R22B	R29B	R30A	R32B	R32B	R33A	R33B	R35B‡	R36A	R36B	R37A	160	30
				50	R17B	R20B	R23B	R30A	R31A	....	R33A	R34A	....	R36A	R37A	R37A	R38A	120	26
				25	R19B	R22B	R25B	R32B	R33A	....	R35B	R36A	....	R38B	R39A	R39B	R39C	65	18
0	BH & BR B	5-8	1861-S1	60	R16A	R19B	R22B	R29B	R30A	R32B	R32B	R33A	R33B	R35B‡	R36A	R36B	R37A	160	30
				50	R17B	R20B	R23B	R30A	R31A	....	R33A	R34A	....	R36A	R37A	R37A	R38A	120	26
				25	R18C	R21C	R24C	R31C	R51C	....	R34C	R35B	....	R37C	R38B	R38C	R39C	80	20
0	BH & BR B	All	★1861-S14	60 / 50	115/230-G4 220/380-G14 240/480-G5 / 115/230-G5 220/380-G15 220/440-G8														
0	SB	All	31041-400	60	....	....	20	Δ	42	48	Δ	51	52	Δ	60	Δ	62	245	27
				50			22	42	43	....	51	52	....	60	61	62	64	232	26
1	C & M	All	2936-S1	60	....	....	C19A	C26B	C27A	C29B	C29B	C30A	C30B	C32B‡	C33A	C33B	C34A	170	40
				50			C20A	C27A	C28A	....	C30A	C31A	....	C33A	C34A	C34A	C35A	125	27
				25			C22A	C29B	C30A	....	C32B	C33A	....	C35B	C36A	C36B	C37A	100	24
1	C	All	★2936-S21	60 / 50 / 25	115/230-G4 240/480-G11 / 110/220-G3 220/440-G11 / 115/230-G7 220/440-G10														
8&1P	SC	All	31041-400	60	....	....	20	Δ	42	48	Δ	51	52	Δ	60	Δ	62	245	27
				50			32	42	43	....	51	52	....	60	61	62	64	232	26
0,1&1P	SB & SC	All	★31041-402	60 / 50	115/230-01 120/240-02 240/480-04 / 115/230-03														
2	D & P	2-4	1707-S1	60	....	T10B	T13B	T20A	T21	T23A	T23A	T24	T24A	T26A‡	T26B	T27A	T27B	465	80
				50		T11A	T14A	T21	T21A	....	T24	T24A	....	T26B	T27A	T27B	T28A	400	65
				25		T13A	T16A	T22B	T23A	....	T25B	T26A	....	T28B	T29A	T29B	T30A	340	51
2	SD	2&3	31063-409	60	....	....	16	Δ	38	44	Δ	47	49	Δ	57	Δ	60	311	37
				50			17	38	39	....	47	48	....	57	58	60	61	296	36
2	SD	4&5	31063-400	60	....	....	16	Δ	38	44	Δ	47	49	Δ	57	Δ	60	438	38
				50			17	38	39	....	47	48	....	57	58	60	61	429	37
3	SE	2&3	31074-400	60	....	....	16	Δ	38	44	Δ	47	49	Δ	57	Δ	60	700	46
				50			17	38	39	....	47	48	....	57	58	60	61	678	47
3	SE	4&5	31091-400	60	....	....	16	Δ	38	44	Δ	47	49	Δ	57	Δ	60	1185	85
				50			17	38	39	....	47	48	....	57	58	60	61	1260	89
3	E&Q	2-4	1775-S1	60	....	....	U11A	U17B	U18A	U20D	Δ	U21A	U21B	U23B‡	U24A	U24B	U25A	800	150
				50			U12A	U18A	U19A	....	U21A	U22A	....	U24A	U25A	U25A	U26A	600	120
				25			U14A	U20B	U21A	....	U23B	U24A	....	U26B	U27A	U27B	U28A	450	90
4	SF	All	31091-400	60	....	....	16	Δ	38	44	Δ	47	49	Δ	57	Δ	60	1185	85
				50			17	38	39	....	47	48	....	57	58	60	61	1260	89
4	F (Series C)	2-4	1775-S1	60	....	....	U11A	Δ	U18A	U20D	Δ	U21A	U21B	Δ	U24A	Δ	U25A	1490	140
				50			U12A	U18A	U19A	....	U21A	U22A	....	U24A	U25A	U25A	U26A	1200	110
				25			U14A	U20B	U21A	....	U23B	U24A	....	U26B	U27A	U27B	U28A	840	80
4	F (Series C)	5	1775-S1	60				Δ	U17B	U20	Δ	U20B	....	Δ	U23B	Δ	U24B	1800	160
				50				UI7B	UI8B	....	U20B	U21B	....	U24B	U24B	U24B	U25B	1450	125
				25				U20A	U20B	....	U23A	U23B	....	U26A	U26B	U27A	U27B	900	90
5	SG	All	31096-400	60				Δ	09	15	Δ	18	19	Δ	24	Δ	29	2970	212
				50				09	10	....	18	....	24	....	29	30		2970	250
5	G & X (Series B)	All	2938-S1	60				Δ	F14A	F16D	Δ	F17A	F17B	Δ	F20A	Δ	F21A	2800	290
				50				F14A	F14B	....	F17A	F17B	....	F20A	F21A	F21A	F21B	2000	220
				25				F16B	F17A	....	F19B	F20A	....	F22B	F23A	F24	F24A	1550	160

FIGURE 8-13   Magnet coil data (*Courtesy* Square D Company).

specification numbers. Similar tables are available from other manufacturers.

Magnet coil voltage ratings. The minimum control voltage that will cause the armature to start to move is called the *pick-up voltage.* The *seal-in voltage* is the minimum control voltage required to cause the armature to seat. Contact life is extended, and contact damage under abnormal voltage conditions is reduced, if the pick-up voltage is also sufficient for seal in. If the voltage is reduced sufficiently, the controller will open. The voltage at which this happens is called the *drop-out voltage.* Drop-out voltage is somewhat less than seal-in voltage.

Magnet coil voltage variations. Note that the voltage applied to the magnet coil is called the *control voltage* and is separate from the *power voltage* applied through the contacts to the motor.

NEMA standards require that the magnetic device operate properly at varying control voltages, from a high of 110 per cent to a low of 85 per cent of rated coil voltage. This range, established by coil design, ensures that the coil will withstand given temperature rises at voltages up to 10 per cent over rated voltage and that the armature will pick up and seal in, even though the voltage may drop to 15 per cent under the nominal rating.

If the voltage applied to the coil is too high, the coil will draw more than its design current. Excessive heat will be produced and will cause early failure of the coil insulation. The magnetic pull will be too high, which will cause the armature to slam home with excessive force.   The magnet faces will wear rapidly, leading to a shortened life for the controller. In addition, contact "bounce" may be excessive, resulting in reduced contact life.

Low control voltages produce low coil currents and reduced magnetic pull. This can result in poor contact closure, chattering of the contacts, arcing, and possible damage to the contacts and coil.

Note that any magnetic device, including magnetic starter coils, produce a characteristic hum. The hum or noise is due mainly to the changing magnetic pull, inducing mechanical vibration. When there is excessive hum, this indicates that the contactor, starter, or relay is defective.

8-5.3 Power circuits for magnetic control devices

The power circuits of a magnetic control device carry the power from the distribution system to the motor. The power circuits and

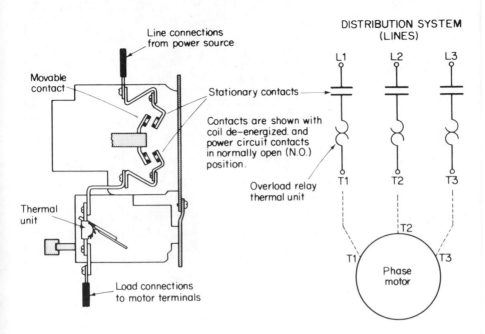

FIGURE 8-14   Power circuits and internal construction of magnetic starter (*Courtesy* **Square D Company**).

internal construction of a magnetic starter are shown in Fig. 8-14. As shown, the power circuits of the starter include the stationary and movable contacts and the thermal unit or heater portion of the overload relay assembly. The number of contacts (or "poles") is determined by the electrical service. In a three-phase three-wire system, for example, a three-pole starter is required.

From a design standpoint, there are two major concerns in the power circuits of motors. First, the conductors (or lines) must be sized on the basis of motor current. Generally, the FLC of the motor is used to determine conductor size, voltage drop, and so on, following the procedures of Chapter 3. The overcurrent devices for the power wiring branch circuit must also be based on the same currents, as discussed in Chapter 3 and in Section 8-3.1.

Second, the power circuit contacts of the controller or starter must be sized on the basis of motor current. This problem is simplified by means of tables that show electrical ratings of magnetic contactors and starters. Such a table is shown in Fig. 8-15. This table is for equipment manufactured by the Square D Company. However, the ratings are related directly to NEMA sizes. The ability of the

NEMA Size	Volts	Max HP Nonplugging & Nonjogging Duty — Single Phase	— Poly Phase	Max HP Plugging & Jogging Duty † — Single Phase	— Poly Phase	Continuous Current Rating, Amperes—600 Volt Max	Service-Limit Current Rating, Amperes *	Tungsten & Infrared Lamp Load, Amperes—250 Volts Max ★	Resistance Heating Loads, KW—other than Infrared Lamp Loads ‡ — Single Phase	— Poly Phase	KVA Rating for Switching Transformer Primaries at 50 or 60 Cycles ▲ — Single Phase	— Poly Phase	3 Phase Rating for Switching Capacitors ◑ Kvar
00	115	1/3	...	...	...	9	11	5	...	...	...	...	...
	200	...	1 ½	...	...	9	11	5	...	...	...	...	...
	230	1	1 ½	...	...	9	11	5	...	...	...	...	...
	380	...	1 ½	...	...	9	11	...	...	...	...	...	...
	460	...	2	...	...	9	11	...	...	...	...	...	...
	575	...	2	...	...	9	11	...	...	...	...	...	...
0	115	1	...	½	...	18	21	10	...	...	...	1.2	...
	200	...	3	...	1 ½	18	21	10	...	...	0.9	1.4	...
	230	2	3	1	1 ½	18	21	10	...	...	1.4	1.7	...
	380	...	5	...	1 ½	18	21	...	...	...	...	2.0	...
	460	...	5	...	2	18	21	...	...	...	1.9	2.5	...
	575	...	5	...	2	18	21	...	...	...	1.9	2.5	...
1	115	2	...	1	...	27	32	15	3	5	1.4	1.7	...
	200	...	7 ½	...	3	27	35	15	...	9.1	...	3.5	...
	230	3	7 ½	2	3	27	35	15	6	10	1.9	4.1	...
	380	...	10	...	5	27	35	...	...	16.5	...	4.3	...
	460	...	10	...	5	27	35	...	12	20	3	5.3	...
	575	...	10	...	5	27	32	...	15	25	3	5.3	...
1P	115	3	...	1 ½	...	36	42	24	...	...	...	...	...
	230	5	...	3	...	36	42	24	...	...	...	...	...
2	115	3	...	2	...	45	52	30	5	8.5	1.9	4.1	...
	200	...	10	...	7 ½	45	52	30	...	15.4	...	6.6	11.3
	230	7 ½	15	5	10	45	52	30	10	17	4.6	7.6	13
	380	...	25	...	15	45	52	...	...	28	...	9.9	21
	460	...	25	...	15	45	52	...	20	34	5.7	12	26
	575	...	25	...	15	45	52	...	25	43	5.7	12	33
3	115	7 ½	...	...	...	90	104	60	10	17	4.6	7.6	...
	200	...	25	...	15	90	104	60	...	31	...	13	23.4
	230	15	30	...	20	90	104	60	20	34	8.6	15	27
	380	...	50	...	30	90	104	...	...	56	...	19	43.7
	460	...	50	...	30	90	104	...	40	68	14	23	53
	575	...	50	...	30	90	104	...	50	86	14	23	67
4	200	...	40	...	25	135	156	120	...	45	...	20	34
	230	...	50	...	30	135	156	120	30	52	11	23	40
	380	...	75	...	50	135	156	...	...	86.7	...	38	66
	460	...	100	...	60	135	156	...	60	105	22	46	80
	575	...	100	...	60	135	156	...	75	130	22	46	100
5	200	...	75	...	60	270	311	240	...	91	...	40	69
	230	...	100	...	75	270	311	240	60	105	28	46	80
	380	...	150	...	125	270	311	...	...	173	...	75	132
	460	...	200	...	150	270	311	...	120	210	40	91	160
	575	...	200	...	150	270	311	...	150	260	40	91	200
6	200	...	150	...	125	540	621	480	...	182	...	79	139
	230	...	200	...	150	540	621	480	120	210	57	91	160
	380	...	300	...	250	540	621	...	...	342	...	148	264
	460	...	400	...	300	540	621	...	240	415	86	180	320
	575	...	400	...	300	540	621	...	300	515	86	180	400
7	230	...	300	...	...	810	932	720	180	315	...	...	240
	460	...	600	...	...	810	932	...	360	625	...	...	480
	575	...	600	...	...	810	932	...	450	775	...	...	600
8	230	...	450	...	...	1215	1400	1080	...	...	...	...	360
	460	...	900	...	...	1215	1400	...	...	...	...	...	720
	575	...	900	...	...	1215	1400	...	...	...	...	...	900

Tables and footnotes are taken from NEMA Standards Publication No. IC 1–1965 Section 2, Part 1 for Magnetic Contractors and Section 3, Parts 21B, 21C, 21D and 21F for Magnetic Starters and includes 1971 revisions for 200 V. and 380 V. ratings.

† Ratings shown are for applications requiring repeated interruption of stalled motor current or repeated closing of high transient currents encountered in rapid motor reversal, involving more than five openings per minute such as plug-stop, plug-reverse or jogging duty. Ratings apply to single speed and multi-speed controllers.

* Per NEMA Standards paragraph IC 1–21 A.20, the service-limit current represents the maximum rms current, in amperes, which the controller may be expected to carry for protracted periods in normal service. At service-limit current ratings, temperature rises may exceed those obtained by testing the controller at its continuous current rating. The ultimate trip current of over-current (overload) relays or other motor protective devices shall not exceed the service-limit current ratings of the controller.

★ FLUORESCENT LAMP LOADS—300 VOLTS AND LESS— The characteristics of fluorescent lamps are such that it is not necessary to derate Class 8502 contactors below their normal continuous current rating. Class 8903 contractors may also be used with fluorescent lamp loads. For controlling tungsten and infrared lamp loads, Class 8903 ac lighting contactors are recommended. These contactors are specifically designed for such loads and are applied to their full rating as listed in the Class 8903 Section. Do not use Class 8903 contactors with motor loads or resistance heating loads.

‡ Ratings apply to contactors which are employed to switch the load at the utilization voltage of the heat producing element with a duty which requires continous operation of not more than five openings per minute

▲ Applies to contactors used with transformers having an inrush of not more than 20 times their rated full load current, irrespective of the nature of the secondary load.

◑ Kilovar ratings of contactors employed to switch power capacitor loads. When capacitors are connected directly across the terminals of an alternating current motor for power factor correction, the motor manufacturer should be consulted as to the maximum size of the capacitor and the proper rating of the motor overcurrent protective device.

"CAUTION: For three phase motors having locked-rotor KVA per horsepower in excess of that for the motor code letters in the right table, do not apply the controller at its maximum rating without consulting the factory. In most cases, the next higher horsepower rated controller should be used."

Controller HP Rating	Maximum Allowable Motor Code Letter
1½–2	L
3–5	K
½ & above	H

FIGURE 8-15 Electrical ratings for AC magnetic contactors and starters (*Courtesy* Square D Company).

starter contacts to carry the FLC without exceeding a rated temperature rise and the isolation from adjacent contacts corresponds to NEMA standards established to categorize the NEMA size of the starter. The starter must also be capable of interrupting the motor circuit under LRC conditions.

Basic selection procedure. Once the motor has been selected and the power wiring sized, the next step is to select the correct starter or controller size. To be suitable for a given motor application, the magnetic starter selected should equal or exceed the motor horsepower and FLC ratings.

For example, assume that the motor to be controlled has a 50-hp rating, the service is 230 V, polyphase, and the motor FLC is 125 A.

By reference to the table of Fig. 8-15 it will be seen that a NEMA size 4 starter would be required for normal motor duty. If the motor is used for jogging or plugging duty, a NEMA size 5 starter should be chosen.

For a NEMA size 4 at 230 V, the maximum polyphase horsepower rating is 50 (normal duty), and 30 (plugging and jogging duty). The continuous current rating is 135 A, which is 10 A higher than the required 125 A.

For a NEMA size 5 at 230 V, the maximum polyphase horsepower rating is 100 (normal duty), and 75 (plugging and jogging duty). The continuous current rating is 270 A, well beyond the required 125 A.

Other information. The table of Fig. 8-15 also provides additional information that can be used in design of electric motor systems, as well as other devices. This information is covered in notes at the bottom of the table. The following explanations supplement these notes.

The *service limit current ratings* show the absolute maximum current that should be provided by overcurrent relays or other motor protective devices in the system. For example, for a NEMA size 4 at 230 V, the motor controlled by the starter should be provided with circuit breakers (or other overcurrent devices) that do not exceed 156 A.

The *tungsten and infrared lamp load ratings* show the maximum current that can be handled by the starter when the starter is used to control lighting circuits (tungsten or infrared lamps only). Refer to Sections 6-4, and 8-7. For example, a NEMA size 4 (at 230 V) can be based to control lighting loads up to 120 A.

The *resistance heating load ratings* show the maximum loads that can be handled by the starter when the starter is used to control resistance heating elements (but not infrared lamps used as heating elements). For example, a NEMA size 4 (at 230 V) can be used to control heating loads of 30 kW (single-phase) or 52 kW (three-phase).

The *transformer ratings* show the maximum kVA ratings when the starters are used to control transformer circuits (that is, when the starters are used to switch transformer loads). For example, a NEMA size 4 (at 230 V) can be used to switch 11-kVA transformers (single-phase) or 23-kVA transformers (three-phase).

The *capacitor ratings* show the kvar ratings when starters are used to control circuits that have power factor correction capacitors. The use of such capacitors is described in Section 9-13. Keep in mind that the ratings are for the contactor or starter, not the capacitor. For example, a NEMA size 4 (at 230 V) can be used to switch circuits with a 40-kvar capacitor. If a larger capacitor is used, say a 75 kvar, it is necessary to use a larger-size controller or starter (NEMA size 5 in this case).

8-5.4 Control circuits for magnetic control devices

The control circuits of a magnetic control device carry the current from the start switch, push button, pilot device, and so on, to the magnetic coil of the controller. This circuit to the magnet coil, which causes the starter or controller to pick up and drop out, is distinct from the power circuit (Section 8-5.3). Although the power circuit can be single-phase or polyphase, the coil circuit is always a single-phase circuit. Elements of a coil circuit include the following:

1. The magnet coil.
2. The contacts of the overload relay assembly.
3. A momentary or maintained contact pilot device, such as a push-button station, pressure, temperature, liquid level, or limit switch.
4. In lieu of a pilot device, the contacts of a relay or timer.
5. An auxiliary contact on the starter, designated as a *holding circuit interlock*, which is required in certain control schemes.

The coil circuit is generally identified as the *control circuit* and contacts in the control circuit handle the *coil load*.

The wiring diagram shown in Fig. 8-16a covers the control circuit wiring provided at the factory in a typical magnetic starter. Per NEMA standards, the single-phase control circuit is conventionally wired be-

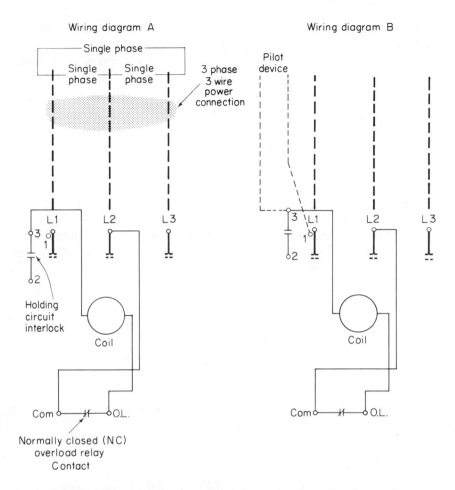

FIGURE 8-16   Typical magnetic starter wiring (*Courtesy* Square D Company).

tween lines 1 and 2. As shown in Fig. 8-16a, the control circuit is connected to the single-phase circuit at line 2, but there is no control circuit connection to line 1.

The diagram of Fig. 8-16b shows that the control circuit is completed by the additional wiring of a pilot device between terminal 3, on the auxiliary contact, and terminal 1 (line 1) on the starter.

Control circuit currents. Although the power circuit and control circuit voltage may be the same, the current drawn by the motor in the power circuit is much higher than that drawn by the coil in the control circuit. Pilot devices and contacts of timers and relays used in control circuits are thus not generally horsepower rated, and the current rating is low compared to a starter or contactor.

Inrush and sealed currents of a control circuit can be determined by reference to a magnet coil table such as shown in Fig. 8-13. For example, a standard-duty push button with a rating of 15 A inrush, 1.5 A normal (sealed) current at 240 V, 60 Hz, can satisfactorily be used to control the coil circuit of a three-pole NEMA size 3 starter or contactor, which has an inrush current of 2.9 A (700 VA/240 V) and a sealed current of 0.2 A (46 VA/240 V).

As a comparison of the differences in current, the *power circuit* contacts of the same starter may be controlling a 30-hp polyphase motor, drawing a FLC of 78 A.

From a design standpoint, the major concern in control circuit wiring is that the conductors must be sized on the basis of coil current, not motor current. Generally, the sealed-in current is used to determine conductor size, voltage drop, and so on, following the procedures of Chapter 3.

Keep in mind that there is some voltage drop between the coil and the pilot device. This is dependent upon conductor resistance, length of the conductor, and coil current. If the voltage should drop below the pick-up or seal-in voltage, it is possible that the magnetic control contacts will not close when the pilot device is operated. Generally, this is a result of *too small conductors* for a given coil current and length of conductor run.

8-5.5 Two-wire control

Figure 8-17 shows the wiring diagrams for the power circuit and control circuit of a three-phase motor control system. The circuit shown is called a two-wire system, since only two wires (or conductors) are used between the control device (start switch, push button, etc.) and the magnetic starter. When the contacts of the control device close, they complete the coil circuit of the starter, causing it to pick up and connect the motor to the lines. When the control device contacts open, the starter is deenergized, stopping the motor.

Two-wire control provides low-voltage release but not low-voltage protection. Wired as shown in Fig. 8-17, the starter will function automatically in response to the direction of the control device, without the attention of an operator.

The dotted portion shown in the elementary diagram represents the *holding circuit interlock* furnished on the starter but not used in

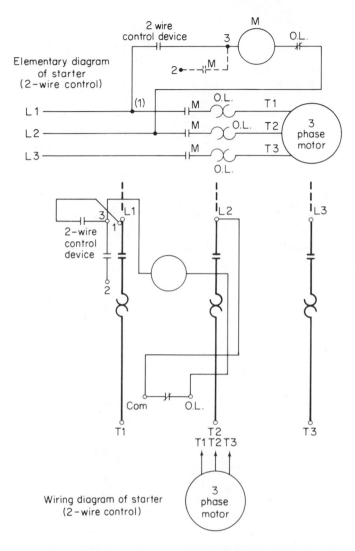

Elementary diagram
of starter
(2-wire control)

Wiring diagram of starter
(2-wire control)

FIGURE 8-17   Two wire control (*Courtesy* Square D Company).

two-wire control.   For greater simplicity, this portion is omitted
from the conventional two-wire elementary diagram.

8-5.6 Three-wire control

Figure 8-18 shows the wiring diagrams for the power circuit and
control circuit of a three-phase motor control system, using three-wire
control (three wires between control device and magnetic starter).

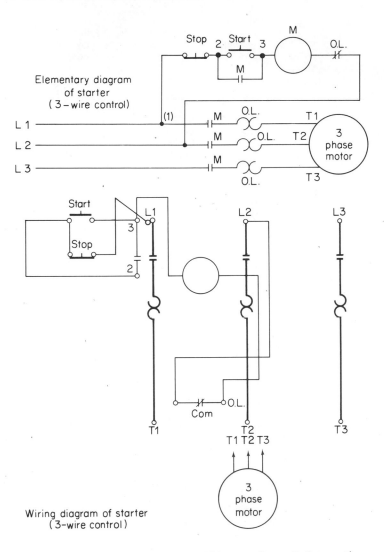

FIGURE 8-18   Three wire control (*Courtesy* Square D Company).

A three-wire control circuit uses momentary contact start–stop buttons and a holding circuit interlock wired in parallel with the start button to maintain the circuit.

Pressing the normally open start button completes the circuit to the coil. The power circuit contacts in lines 1, 2, and 3 close, completing the circuit to the motor, and the holding circuit contacts (mechanically linked with the power contacts) also closes. Once the starter has picked up, the start button can be released, as the now-closed interlock contact provides an alternative current path around the reopened start contact.

Pressing the normally closed stop button will open the circuit to the coil, causing the starter to drop out. An overload condition, which causes the overload contact to open, a power failure, or a drop in voltage to less than the seal-in value will also deenergize the starter. When the starter drops out, the interlock contact reopens, and both current paths to the coil (through the start button and the interlock) are now open.

8-5.7 Holding circuit interlock

The holding circuit interlock is a normally open auxiliary contact provided on standard magnetic starters and contactors. The interlock closes when the coil is energized to form a holding circuit for the starter after the start button has been released, as described for three-wire control in Section 8-5.6.

In addition to the main or power contacts that carry the motor current, and the holding circuit interlock, a starter can be provided with externally attached auxiliary contacts, commonly called *electrical interlocks*. Interlocks are rated to carry *only control circuit* currents, and not motor (power) currents. Normally open and normally closed versions of electrical interlocks are available.

Among a wide variety of applications, interlocks can be used to control other magnetic devices where sequence operation is desired; to electrically prevent another controller from being energized at the same time (such as the reversing starters, Section 8-7); and to make and break circuits to indicating or alarm devices such as pilot lights, bells, or either signals. Electrical interlocks are packaged in kit form and can be easily added in the field.

8-5.8 Control device (pilot device)

A device that is operated by some nonelectrical means (such as the movement of a lever) and has contacts in the control circuit of a starter is called a *control device*. Operation of the control device will control the starter and hence the motor. Typical control devices are the switches described in Section 8-6.

Some control devices have a horsepower rating and are used to directly control small motors through the operation of their contacts.

When used in this way, separate overload protection (such as a manual starter) normally should be provided, as the control device does not usually include overload protection.

A *maintained-contact* control device is one which when operated will cause a set of contacts to open (or close) and stay open (or closed) until a deliberate reverse operation occurs. A conventional thermostat is a typical maintained-contact device. Maintained-contact control devices are used with two-wire control.

A standard push button is a typical *momentary-contact* control device. Pushing the button will cause normally open contacts to close and normally closed contacts to open. When the button is released, the contacts revert to their original states. Momentary-contact devices are used with three-wire control or jogging service.

8-5.9 Low-voltage (undervoltage) release

A two-wire control provides *low-voltage release*. The term describes a condition in which a reduction or loss of voltage will stop the motor but in which motor operation will automatically resume as soon as power is restored.

If the two-wire control device in the diagram of Fig. 8-17 is closed, a power failure or drop in voltage below the seal-in value will cause the starter to drop out, but as soon as power is restored or the voltage returns to a level high enough to pick up and seal, the starter contacts will reclose and the motor will again run. This is an advantage in applications involving unattended pumps, refrigeration processes, ventilating fans, etc.

However, in many applications, the unexpected restarting of a motor after power failure is undesirable. If protection from the effects of a low-voltage condition is required, the three-wire system should be used.

8-5.10 Low-voltage (undervoltage) protection

A three-wire control system such as the one shown in Fig. 8-18 provides *low-voltage protection*, in contrast to the low-voltage release of the two-wire system. In either system the starter will drop out and the motor will stop in response to a low-voltage condition or power failure.

When power is restored, the starter connected for three-wire control will not pick up, as the reopened holding circuit contact and the normally open start button contact prevent current flow to the coil. To restart the motor after a power failure, the low-voltage protection offered by a three-wire control requires that the start button be pressed.   A deliberate action must be performed, ensuring greater safety than that provided by two-wire control.

8-5.11 Full voltage (across-the-line) starter

As the name implies, a *full voltage*, or *across-the-line*, *starter* directly connects the motor to the line. All the starters described in this chapter are full voltage starters.

A motor connected across-the-line draws full inrush current and develops maximum torque so that it accelerates the load to full speed in the shortest possible time.   Across-the-line starting can be used wherever this high inrush current and starting torque are not objectionable.

With some loads the high starting torque and high inrush current are not acceptable. In those cases, reduced voltage starting is used. The wiring and equipment for such control systems are highly specialized and are not covered in this book.

8-5.12 Common control versus transformers and separate control

The coil circuit of a magnetic starter or contactor is distinct from the power circuit. The coil circuit can be connected to any single-phase power source, provided that the *coil voltage and frequency* match service to which it is connected.

When the control circuit is tied back to lines 1 and 2 of the starter, the voltage of the control circuit is always the same as the power circuit voltage, and the term *common control* is used.

It is sometimes desirable to operate push buttons or other control circuit devices at some voltage *lower* than the motor voltage. In Fig. 8-19 a single-phase control transformer (with dual voltage 240/480-V primary and 120-V secondary) has its 480-V primary connected to the 480-V three-phase three-wire service brought into the starter. The control circuit is supplied by the 120-V secondary. The coil voltage is 120 V, and the push buttons or other control devices operate at the same voltage level. A fuse is often used to protect the

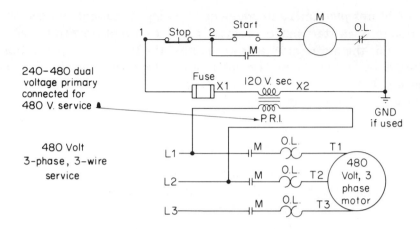

FIGURE 8-19   Wiring for control circuit with transformer (*Courtesy* Square D Company).

control circuit (in addition to branch circuit overcurrent protection), and it is common practice to ground one side of the transformer.

Control of a power circuit by a lower control circuit voltage can be obtained by connecting the coil circuit to a *separate control* voltage source rather than to a transformer secondary. Such a separate control system is shown in Fig. 8-20. The coil rating must match the control source voltage, but the power circuit can be any voltage (typically up to 600 V).

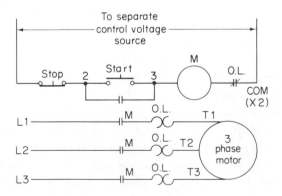

FIGURE 8-20   Wiring for control circuit supplied by separate source (separate control) (*Courtesy* Square D Company).

8-6 SWITCHES AS CONTROL DEVICES

Switches are the most common control devices used in motor control circuits. The following is a summary of the switch types in most common use.

Drum switch. The *drum switch* is a manually operated three-position three-pole switch used for *manual reversing* of single-phase and three-phase motors. Drum switches are often rated by motor horsepower, in addition to voltage and current.

Control station. A *control station* (also known as a *push-button station*) may contain push buttons, selector switches, and pilot lights. Figures 8-21 through 8-23 show some typical control stations. Push buttons may be of the momentary- or maintained-contact type. Selector switches are usually maintained contact or can be of the spring-return type to give momentary-contact operation.

FIGURE 8-21 Standard duty push-button (or control) station (*Courtesy* Square D Company).

Standard duty stations (Fig. 8-21) will handle the coil currents of contactors up to NEMA size 4. Heavy-duty stations (Figs. 8-22 and 8-23) have higher contact ratings and provide greater flexibility through a wider variety of controls and interchangeability of units.

Foot switch. A *foot switch* is a control device operated by a foot pedal used where the process or machine requires that the operator have both hands free. Foot switches can be of momentary-contact type or mechanically latched.

Limit switch. A *limit switch* is a control device that converts mechanical motion into an electrical control signal. Typical appli-

FIGURE **8-22   Heavy-duty pushbutton (control) station** (*Courtesy* **Square D Company**).

cations include start, stop, reverse, slow down, speed up, or recycle machine operations.

Snap switch. S*nap switches* for motor control purposes are enclosed precision switches that require low operating forces and have a high repeat accuracy.

FIGURE 8-23   Heavy-duty oiltight pushbutton (control) station (*Courtesy* Square D Company).

Pressure switch.   A *pressure switch* is a control device for pumps, air compressors, welding machines, lubrication systems, and machine tools.   Pressure switch contacts are operated by pistons, bellows, or diaphragms against a set of springs.   The spring pressure determines the pressure at which the switch closes and opens its contacts.

Float switch.   A *float switch* is a control device whose contacts are controlled by movement of a rod or chain and a counterweight fitted with a float.   Float switches are often used for tank pump motors.

8-7 CONTACTORS

As discussed, *contactors* are magnetic control devices used where high currents are involved.   The contacts can be held electrically or mechanically.   In a conventional contactor, current flow through the coil creates a magnetic pull to seal in the armature and maintain the contacts in a switched position.   (That is, normally open contacts will be held closed, and normally closed contacts will be held open).   Because the contactor action is dependent on the current flow through the coil, the contactor is described as *electrically held.* As soon as the coil is deenergized, the contacts will revert to their initial position.

Mechanically held versions of contactors (and relays, Section 8-8) are also available.   The action is accomplished through use of two coils and a latching mechanism.   Energizing one coil (latch coil) through a momentary signal causes the contacts to switch, and a mechanical latch holds the contacts in this position, even though the initiating signal is removed, and the coil is deenergized.   To restore the contacts to their initial position, a second coil (unlatch coil) is momentarily energized.   Mechanically held contactors and relays are used where the slight hum of an electrically held device would be objectionable, as in auditoriums, hospitals, and churches.

Two typical applications for contactors are as a reversing starter for an electric motor and as a controller for light systems.

8-7.1 Reversing starter

Reversing the direction of motor shaft rotation is often required. Three-phase squirrel-cage motors can be reversed by *reconnecting any two of the three line connections to the motor.* By interwiring two contactors, an electromagnetic method of making the reconnection can be obtained. A NEMA size 1 three-pole reversing starter is shown in Fig. 8-24. The wiring diagram is shown in Fig. 8-25.

The contacts (F) of the forward contactor, when closed, connect lines 1, 2 and 3 to motor terminals T1, T2, and T3, respectively. As long as the forward contacts are closed, mechanical and electrical interlocks prevent the reverse contactor from being energized.

When the forward contactor is deenergized, the second contactor can be picked up, closing its contacts (R), which reconnect the lines to the motor. Note that by running through the reverse contacts, line 1 is connected to motor terminal T3 and line 3 is connected to motor terminal T1. The motor will now run in the opposite direction.

Whether operating through either the forward or reverse contactor, the power connections are run through an overload relay as-

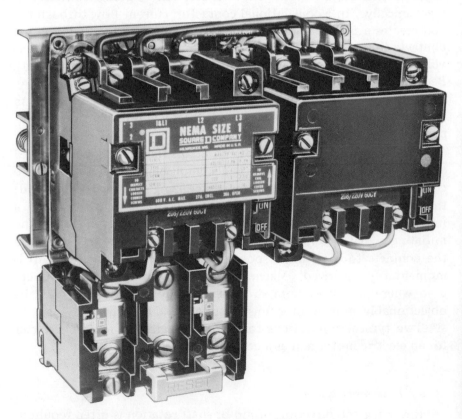

FIGURE 8-24   NEMA Size 1 reversing starter (*Courtesy* Square D Company).

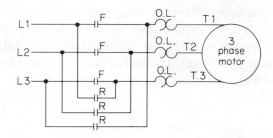

FIGURE 8-25   Wiring diagram for reversing starter (*Courtesy* Square D Company).

sembly, which provides motor overload protection. A magnetic reversing starter consists of a starter and contactor (suitably interwired) with electrical and mechanical interlocking to prevent the coil of both units from being energized at the same time.

Manual reversing starters (using two manual starters) are also available. As in the magnetic version, the forward and reverse switching mechanisms are mechanically interlocked, but since coils are not used in the manually operated equipments, electrical interlocks are not furnished.

8-7.2 Lighting contactor

The contactors used for the control of electric motors can also be used to control lighting systems, as shown in Fig. 8-15. However, standard motor control contactors must be derated if used to control lighting loads, to prevent welding of contacts on the high initial current. Filament-type lamps have inrush currents approximately 15 to 17 times the normal operating currents.

As shown in Fig. 8-15, a NEMA size 1 contactor has a continuous current rating of 27 A, but if it is used to switch certain lighting loads, it must be derated to 15A. However, the standard contactor need not be derated for resistance heating or fluorescent lamp loads, which do not impose as high an inrush current.

Special *lighting contactors* differ from standard contactors in that the contact-tip material is a *silver tungsten carbide* which resists welding on high initial currents. A holding circuit interlock is not normally provided, because this type of contactor is frequently controlled by a two-wire (Section 8-5.5) pilot device such as a time clock or photoelectric relay.

Unlike standard contactors, lighting contactors are not horsepower rated or categorized by NEMA size. Instead, lighting contactors are designated by *ampere ratings* (20, 30, 60, 100, 200, and 300 A). It should be noted that lighting contactors are specialized in their application and *should not be used on motor loads*.

8-8 CONTROL RELAYS

A relay is an electromagnetic device whose contacts are used in control circuits of magnetic starters, contactors, solenoids, timers, and other relays. Relays are generally used to *amplify* the contact capability or *multiply* the switching function of a pilot device.

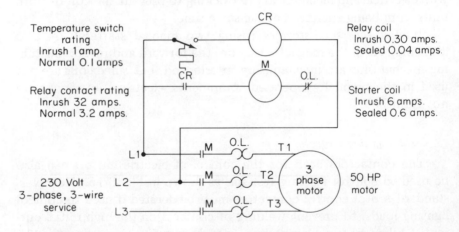

FIGURE 8-26   Relay used for current amplification (*Courtesy* Square D Company).

Figure 8-26 shows an example of *current amplification.* Relay and starter coil voltages are the same (230 V), but the ampere rating of the temperature switch is too low to handle the current drawn by the starter coil M. A relay is interposed between the temperature switch and the starter coil. The current drawn by the relay coil CR is within the rating of the temperature switch, and the relay contact CR has a rating adequate for the current drawn by the starter coil.

Figure 8-27 shows an example of *voltage amplification.* A condition may exist in which the voltage rating of the temperature switch

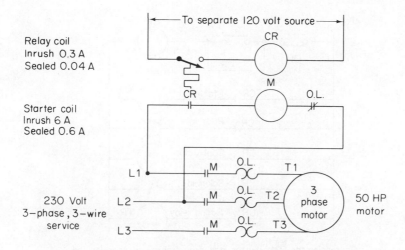

Relay coil
Inrush 0.3 A
Sealed 0.04 A

Starter coil
Inrush 6 A
Sealed 0.6 A

230 Volt
3-phase, 3-wire
service

FIGURE 8-27   Relay for voltage amplification (*Courtesy* Square D Company).

is too low to permit its direct use in a starter control circuit operating at some higher voltage. In this application the coil of the interposing relay and the pilot device are wired to a low-voltage source of power compatible with the rating of the pilot device. The relay contact, with its higher voltage rating, is then used to control operation of the starter.

Figure 8-28 shows an example of *multiplying switching functions* of a pilot device with a single or limited number of contacts. As shown, a single-pole push-button contact can control operation of many different loads (pilot light, starter, contactor, solenoid, and timing relay), through the use of a six-pole relay. Relays are commonly used in complex controllers to provide the logic or "brains" to set up and initiate the proper sequencing and control of a number of interrelated operations.

Relays used in motor control differ in voltage ratings (typically 150, 300, and 600 V), number of contacts, contact convertability, physical size, and in attachments to provide accessory functions such as mechanical latching and timing.

In selecting a relay for a particular application, one of the first steps should be to determine the *control voltage* at which the relay will operate. Once the voltage is known, the relays that have the necessary *contact rating* can be further reviewed and a selection made

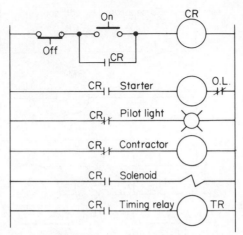

Depressing the "ON" button in this control circuit energies the
relay coil (CR). Its normally open contacts to complete the
control circuits to the starter, solenoid and timing relay, and one
contact forms a holding circuit around the "ON" button. The
normally closed contacts open to de-energize the contactor
and turn off the pilot light.

FIGURE 8-28   Relay used to multiply switching functions (*Courtesy*
Square D Company).

on the basis of *number of contacts* and any other specialized characteristics needed. Figure 8-29 through 8-31 show typical relays used
to motor control, as well as lighting and heating applications.

A *timing relay* is similar to a control relay except that certain of
its contacts are designed to operate at a preset time interval, *after*
the coil is energized or deenergized. A delay on energization is called
on delay. A delay on deenergization is called *off delay*.

FIGURE 8-29   150-volt relay (*Courtesy*
Square D Company).

FIGURE 8-30   300-volt   relay
(*Courtesy* Square D Company).

FIGURE 8-31   600-volt   relay
(*Courtesy* Square D Company).

8-9 COMBINATION STARTERS

In some applications combination starters are used to control electric motors. A combination starter is so named since it combines a disconnect means, which might incorporate a short-circuit protective device (for overcurrent protection, Section 8-3.1) and a magnetic starter, in one enclosure.

Compared with a separately mounted disconnect and starter, the combination starter takes up less space, requires less time to install and wire, and provides greater safety. Personnel safety is assured because the door is *mechanically interlocked*, so that it cannot be opened without first opening the disconnect. Combination starters can be furnished with circuit breakers or fuses to provide overcurrent protection and are available in nonreversing and reversing versions.

Chapter 9

Basic
Electrical Data

This chapter provides a review, or summary, of basic electrical data most needed for design of electrical wiring systems. Basic electrical theory is not included. Complete mathematical analysis of design problems is deliberately avoided. Math is used only where absolutely necessary, and then in the simplest possible form (basic arithmetic, graphic solutions, etc.). The data are included for the student (who needs to understand the problems of design) and for the working electrician or contractor (who needs a ready reference of basic equations, calculation techniques, etc.).

9-1 OHM'S LAW FOR DIRECT CURRENT

Ohm's law equations are more widely used than any other in electrical wiring design. The basic Ohm's law equations for direct current are given by

$$E = IR \qquad I = \frac{E}{R} \qquad R = \frac{E}{I} \qquad P = EI$$

where I is current (in A), R is resistance (in Ω), E is voltage across R (in V), and P is power (in W). The calculations for direct-current Ohm's law are shown in Fig. 9-1.

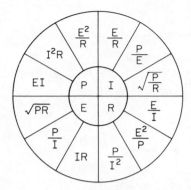

Solution for quantities printed in center section
is calculated by substituting known values in the
equations adjoining the respective center
sections.
For example, if desired quantity is R, and both
E and P are known, $R = E^2/P$

FIGURE 9-1   Direct current Ohm's law equations.

9-2 RESISTANCE IN CIRCUITS

Resistances, either in conductors or in the load, function to control the flow of electrical current and to divide voltages. Although resistances can be used in a variety of circuit combinations, these circuits are versions of the basic series and parallel arrangements shown in Fig. 9-2.

9-2.1 Finding resistance values by means of current

In practical applications it is often convenient to find resistance values by means of currents passing through a circuit. This is particularly true for parallel-resistance networks, since the same voltage appears across each of the resistors. Figure 9-3 shows the equations necessary to find either current or resistance in a parallel-resistance network. This applies where there are two resistors and three of the four values are known. The four equations of Fig. 9-3 are based on what is sometimes referred to the *shunt law*. (The ratio of currents is inversely proportional to the ratio of resistances.)

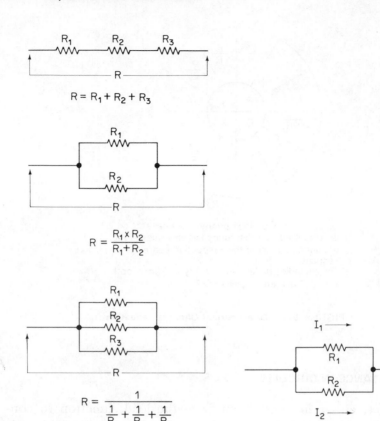

$$R = R_1 + R_2 + R_3$$

$$R = \frac{R_1 \times R_2}{R_1 + R_2}$$

$$R = \frac{1}{\dfrac{1}{R_1} + \dfrac{1}{R_2} + \dfrac{1}{R_3}}$$

or

$$R = \frac{R_1}{1 + \dfrac{R_1}{R_2} + \dfrac{R_1}{R_3}}$$

$$I_1 = \frac{I_2 \times R_2}{R_1} \qquad I_2 = \frac{I_1 \times R_1}{R_2}$$

$$R_1 = \frac{I_2 \times R_2}{I_1} \qquad R_2 = \frac{I_1 \times R_1}{I_2}$$

FIGURE 9-2   Resistance in conductors and circuits.

FIGURE 9-3   Finding resistance values by means of currents.

9-3 KIRCHHOFF'S VOLTAGE LAW

In many electrical circuits the arrangement of the devices and applied voltages makes it almost impossible to solve such networks by simple application of Ohm's law. These problems can often be solved by applying Kirchhoff's voltage law. This law can be stated in many ways:

In any complete electrical circuit, the sum of all the voltage (IR)

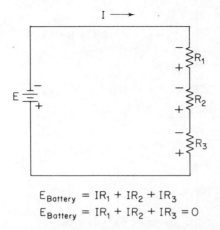

$$E_{Battery} = IR_1 + IR_2 + IR_3$$
$$E_{Battery} = IR_1 + IR_2 + IR_3 = 0$$

FIGURE 9-4   Relationship of voltages and current in a typical series circuit, showing Kirchhoff's voltage law.

drops, taken with their proper signs, is zero; or, the sum of the voltages in a complete circuit is equal to the applied voltage.

Figure 9-4 shows the relationship of the voltages and current in a typical series circuit.

9-3.1 Pointers to help in applying Kirchhoff's voltage law

In the more complicated problems, it is not always easy to determine the direction of current, so simply *assume a direction of current flow.* If the assumption is backwards, the answer for current strength will be numerically correct but will be a *negative* number.

9-3.2 Effect of aiding and opposing voltages in series

When two or more voltage sources are connected in series, their effect is additive if the polarities are aiding. If the polarities are opposed, the smaller voltage is subtracted from the larger. If there are several voltages of various polarities, the total of one polarity is subtracted from the total of the other polarity. Figure 9-5 shows the relationship of the voltages in a typical circuit.

Place *polarity markings* on all voltages and resistors in the circuit. The assumed current direction will not affect the battery polarities, but the voltage drop on resistors will be affected.

By working around the circuit, *set up each term of the equation.*

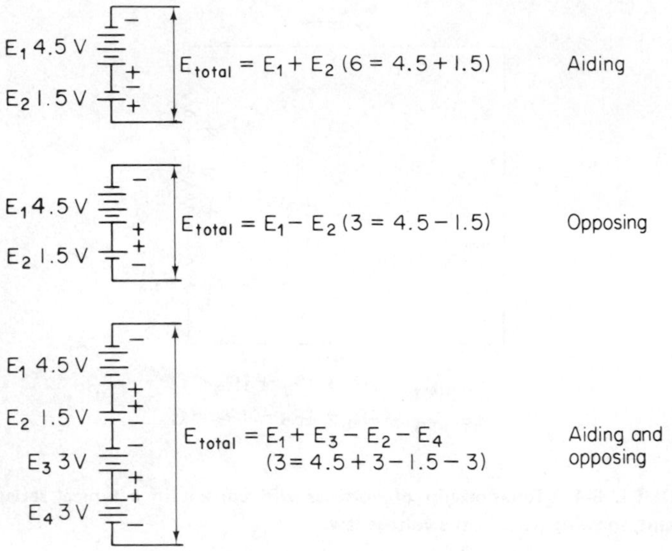

E1 4.5 V
E2 1.5 V
Etotal = E1 + E2 (6 = 4.5 + 1.5)
Aiding
E1 4.5 V
E2 1.5 V
Etotal = E1 − E2 (3 = 4.5 − 1.5)
Opposing
E1 4.5 V
E2 1.5 V
E3 3 V
E4 3 V
Etotal = E1 + E3 − E2 − E4
(3 = 4.5 + 3 − 1.5 − 3)
Aiding and opposing

FIGURE 9-5 **Examples of aiding and opposing voltages in series.**

Include all voltage sources and all voltage drops. In the equation, precede each term by the sign found on *leaving* each particular voltage or resistor.

9-4 KIRCHHOFF'S CURRENT LAW

In some applications it is often more convenient to solve a network by means of Kirchhoff's current law. This law can be stated:

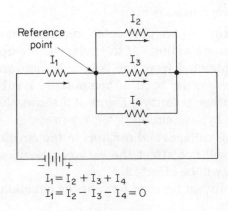

Reference point
I1
I2
I3
I4
I1 = I2 + I3 + I4
I1 = I2 − I3 − I4 = 0

FIGURE 9-6 **Relationship of currents in a series-parallel circuit, showing Kirchhoff's current law.**

In any electrical network the algebraic sum of the currents that meet (entering and leaving) at a point is zero; or the algebraic sum of the currents flowing toward a junction is equal to zero.

Figure 9-6 shows the relationship of the currents in a typical circuit.

9-5 SOLVING RESISTANCE NETWORK PROBLEMS WITH ASSUMED VOLTAGES

It is sometimes convenient to solve resistance network problems with assumed voltages and then to apply Ohm's law. An example of this is shown in Fig. 9-7. Here, a 500-Ω resistance must be placed in parallel with an unknown resistance or load to produce an equivalent 100 Ω.

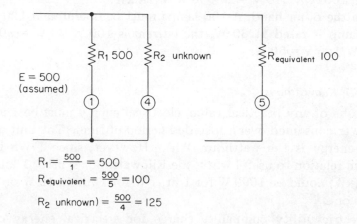

$$R_1 = \frac{500}{1} = 500$$
$$R_{equivalent} = \frac{500}{5} = 100$$
$$R_2 \text{ unknown}) = \frac{500}{4} = 125$$

FIGURE 9-7   Solving resistance network problems with assumed voltages.

If the two resistance values are known and it is desired to find the equivalent resistance, the equation of Fig. 9-2 can be used. However, since it is necessary to find one resistor value, the equation must be rearranged. Instead, assumed voltages can be used as follows:

Assume a voltage of 500 V across both resistors. This will produce 1 A in the known 500-Ω resistor $(I = E/R)$. The same 500 V will produce 5 A through the equivalent 100-Ω resistance. Since 5 A flows through the total network and 1 A flows through the known 500-Ω resistor, 4 A must be flowing through the unknown resistor. Since $R = E/I$, the unknown resistance must be 125 Ω (500/4 = 125).

9-6 BASIC POWER RELATIONSHIPS

A typical electrical system consists of many devices, or loads, of different types and uses. For example, a typical electrical system in an industrial plant might include electric motors, lighting, and heating. It is impractical, if not impossible, to evaluate such loads on the basis of resistance. However, when it is considered that all these loads are in parallel across a particular voltage, either the total current or the total power can be used to sum up the total power consumption.

The relationships are

$$\text{current} = \frac{\text{power}}{\text{voltage}} \qquad \text{power} = \text{current} \times \text{voltage}$$

For example, if 10 electric motors must be operated with 240 V and each motor draws a maximum of 10 A, the power is 10 × 10 A = 100 A; 100 A × 240 V = 24,000 W or 24 kW.

On the other hand, if 100 lamps must be operated at 120 V and each lamp is rated at 60 W, the current is 100 × 60 W = 6000 W; 6000 W/120 V = 50 A.

9-6.1 Kilowatthour

To be of any practical value, electrical energy must be evaluated as power consumed over a specified period of time. The unit of electrical energy is the watthour (Wh). However, since 1 Wh is very small in relation to useful work, the Kilowatthour is used; 1 kilowatthour (kW) could be 1000 W for 1 hr, 1 W for 1000 h, 10 W or 100 h, and so on.

Electric utility companies charge for electrical energy on the basis of kilowatthours. Typically, the charge is from 1 to 10 cents/kW and depends upon such factors as total power consumption, location of facilities, number of meters at the service entrance, demand for a particular period of time, and seasonal conditions.

The relationships are

$$\text{kilowatthour} = \text{power (in kW)} \times \text{time (in hours)}$$

Assume that the 10 electric motors described in the previous example are to be operated for 8 h each day and that the charge is 3 cents/kWh. Find the kWh and the cost per day.

$$\text{kWh} = 24 \times 8 = 192 \qquad \text{cost} = 192 \times 3 = \$5.76$$

9-6.2 Horsepower

Horsepower is the common unit of mechanical power. The watt is the common unit of electric power. The relationships between watts and horsepower are

$$1hp \quad = \quad 746 \text{ W or } 0.746 \text{ kW}$$
$$1kW \quad = \quad 1.34 \text{ hp}$$

Each of the electric motors in the previous example consumed 2.4 kW (240 V $\times$ 10 A). The horsepower rating of each motor is 3.216 (1.34 $\times$ 2.4). However, this presumes that the motor is 100 per cent efficient (not a practical condition).

9-6.3 Power efficiency

Part of the power applied to any circuit or load is consumed by the circuit or load and is not converted directly into energy. The energy or power output from a load will always be less than the power input. The ratio of power output to power input is termed *efficiency* or *power efficiency* and is expressed by

$$\text{efficiency} = \frac{\text{power output}}{\text{power input}}$$

Since output is less than input, the result is a decimal. Multiply the result by 100 to convert into terms of percentage.

For example, assume that our electric motor that has an input of 2.4 kW, or 3.216 hp, produces an output of 3 hp. The efficiency is 0.93, or 93 per cent (3/3.216 = 0.93).

9-7 OHM'S LAW FOR ALTERNATING CURRENT

Ohm's law equations for alternating current are essentially the same as for direct current, except that impedance (Z) is substituted for resistance (R). Therefore,

$$E = \frac{I}{Z} \qquad I = \frac{E}{Z} \qquad Z = \frac{E}{I} \qquad P = \frac{E}{I}$$

where I is the current (in A), Z is the impedance (in Ω), E is the voltage across Z (in V), P is the apparent power (in W). The calculations for alternating-current Ohm's law are shown in Fig. 9-8.

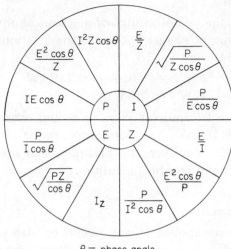

θ = phase angle

Solution for quantities printed in center section is calculated by substituting known values in the equations adjoining the respective center sections.
For example, if desired quantity is Z, and both E and P are known, $Z = E^2 \cos \theta / P$.

FIGURE 9-8   Alternating current Ohm's law equations.

The term "apparent power" is applied when the reactive power factor of an alternating current is disregarded. To obtain true power is an ac circuit it is necessary to multiply the apparent power by the cosine of the phase angle. Since the phase angle is always less than 90° and the corresponding cosine is a fraction of 1, the true power will always be less than the apparent power.

The ratio between the true power and the apparent power is known as the *power factor*. This can be expressed as

$$\text{power factor} = \frac{E \times I \times \text{cosine phase angle}}{E \times I}$$

or

$$\text{power factor} = \text{cosine phase angle} = \frac{R}{Z}$$

where R is the resistance in an ac circuit (in Ω) and, Z is the impedance (in Ω).

In those ac circuits where there is no reactance (pure resistance, such as incandescent lighting), the phase angle is zero, so the apparent power is the true power (the power factor is 1).

9-8 TRIGONOMETRY AND VECTORS

Trigonometry is a branch of mathematics that deals with the properties of triangles. Trigonometry is particularly useful in ac electricity, because some electrical quantities are best expressed as *vectors* or lines that form angular relationships with each other. In some branches of electricity the term "vector" is being replaced by the term *phasor*, which is a rotating vector. A true vector diagram can be made of any factor that has both magnitude and direction (such as wind velocity). Since the direction of alternating current is constantly changing, phasor may be a better term. However, the term "vector" will probably be around for some time to come.

Either way, a rotating vector is simply a diagram of lines drawn to scale in the proper direction to indicate the intensity and direction of the forces to be added or subtracted. For example, suppose that a vector 1 in. long represents a force of 10 V. Then a 2-in. vector line would represent a force of 20 V.

Vectors can be solved by two basic methods: one is the graphic method, where the various lines and angles are laid off on paper using a given scale. The result or solution is then measured physically, using the same scale. The second method involves the use of trigonometry. This method is usually more exact, because it is possible to calculate the length of lines, angles, and so on, rather than to make approximate physical measurements. However, the graphic method is usually simpler and quicker.

9-8.1 Vector concept

Vector diagrams make it possible to add, subtract, multiply, and divide quantities (voltages, currents, etc.) that are *not in phase*. Vectors are the only practical method of combining ac voltages and currents (which are generally out of phase except for pure resistive loads). For best understanding of vectors, it is necessary to visualize a vector diagram in relation to an ac sine wave.

Figure 9-9 is a combination of a sine curve and a vector diagram for an armature rotating in a magnetic field and generating a peak voltage of 10 V.

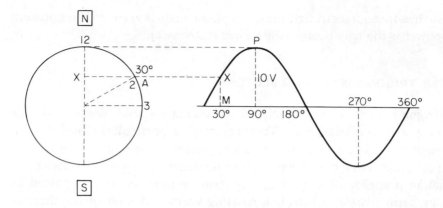

FIGURE 9-9   Combined vector diagram and sine curve.

When the conductor is at 0° (3 position), it is cutting no lines of flux in the magnetic field and is generating no induced voltage. When the conductor moves up to 30° (2 position), an induced voltage of 5 V is generated. Finally, when the conductor reaches 90° (12 position), the peak voltage of 10 V is induced.

If the vector arrow is allowed to rotate along with the indicator and a horizontal line is run from the arrowhead over to the line of peak voltage, such as line AX from the 2 position, the voltage at the 2 position can be found. Measure the length of the line MX. This line is half the length of line OP; OP represents the peak voltage of 10 V, so MX represents 5 V.

9-8.2 Practical vector addition

Assume that an armature has two conductor loops on it—loop B being 30° behind loop A. When this two-loop armature is rotated in the magnetic field, loop A will be cutting a maximum number of lines of flux when loop B is cutting only about three-fourths of the maximum, as shown in Fig. 9-10. Consequently, A will generate peak voltage, while B generates about 75 per cent of peak voltage.

It is possible to combine or add these two voltages with a vector diagram, using physical measurements or trigonometry, whichever is more convenient.

In using physical measurements alone, decide on a convenient scale. Assume that the generator peak voltage is 10 V and allow 1 in. to represent 1 V.

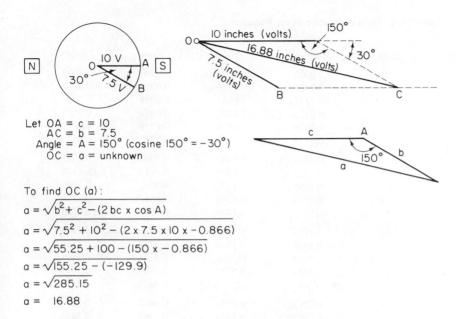

Let OA = c = 10
    AC = b = 7.5
    Angle = A = 150° (cosine 150° = −30°)
    OC = a = unknown

To find OC (a):

$$a = \sqrt{b^2 + c^2 - (2\,bc \times \cos A)}$$

$$a = \sqrt{7.5^2 + 10^2 - (2 \times 7.5 \times 10 \times -0.866)}$$

$$a = \sqrt{55.25 + 100 - (150 \times -0.866)}$$

$$a = \sqrt{155.25 - (-129.9)}$$

$$a = \sqrt{285.15}$$

$$a = \quad 16.88$$

FIGURE 9-10   Vector diagram of two-loop armature generator.

Lay off a line 10 in. long, such as OA of Fig. 9-10. Then lay off a line OB at an angle of 30° to OA but make OB approximately 7.5 in. long to indicate 7.5 V. Next, build up a parallelogram by making AC equal and parallel to OB and BC equal and parallel to OA. Then draw the diagonal OC. Measure the length of OC, which should be approximately $16\frac{7}{8}$ in., of 16.88 V.

The same problem can be solved using trigonometry, as shown in Fig. 9-10. In this case, a rough vector diagram is laid off to construct a parallelogram. The measurements need not be exact. Then the parallelogram is converted to an oblique triangle as indicated, and the problem is solved by use of the equation shown on Fig. 9-10. The cosines of the angles involved are found in tables of trigonometric functions such as those of Table 9-1.

9-8.3 Practical vector subtraction

Vectors can be used to subtract one one-of-phase value from another, as shown in Fig. 9-11. Assume that OE is to be subtracted from OD. Reverse the direction of OE to get a negative (−) OE. Then construct the parallologram. The resultant line OG gives the scale and direction of the difference.

TABLE 9-1 Table of Trigonometric Functions

Angle	Radians	Sine	Cosine	Tangent
0	0.0000	0.0000	1.0000	0.0000
1	0.0175	0.0175	0.9998	0.0175
2	0.0349	0.0349	0.9994	0.0349
3	0.0524	0.0523	0.9986	0.0524
4	0.0698	0.0698	0.9976	0.0699
5	0.0873	0.0872	0.9962	0.0875
6	0.1047	0.1045	0.9945	0.1051
7	0.1222	0.1219	0.9925	0.1228
8	0.1396	0.1392	0.9903	0.1405
9	0.1571	0.1564	0.9877	0.1584
10	0.1745	0.1736	0.9848	0.1763
11	0.1920	0.1908	0.9816	0.1944
12	0.2094	0.2079	0.9781	0.2126
13	0.2269	0.2250	0.9744	0.2309
14	0.2443	0.2419	0.9703	0.2493
15	0.2618	0.2588	0.9659	0.2679
16	0.2793	0.2756	0.9613	0.2867
17	0.2967	0.2924	0.9563	0.3057
18	0.3142	0.3090	0.9511	0.3249
19	0.3316	0.3256	0.9455	0.3443
20	0.3491	0.3420	0.9397	0.3640
21	0.3665	0.3584	0.9336	0.3839
22	0.3840	0.3746	0.9272	0.4040
23	0.4014	0.3907	0.9205	0.4245
24	0.4189	0.4067	0.9135	0.4452
25	0.4363	0.4226	0.9063	0.4663
26	0.4538	0.4384	0.8988	0.4877
27	0.4712	0.4540	0.8910	0.5095
28	0.4887	0.4695	0.8829	0.5317
29	0.5061	0.4848	0.8746	0.5543
30	0.5236	0.5000	0.8660	0.5774
31	0.5411	0.5150	0.8572	0.6009
32	0.5585	0.5299	0.8480	0.6249
33	0.5760	0.5446	0.8387	0.6494
34	0.5934	0.5592	0.8290	0.6745
35	0.6109	0.5736	0.8192	0.7002
36	0.6283	0.5878	0.8090	0.7265
37	0.6458	0.6018	0.7986	0.7536
38	0.6632	0.6157	0.7880	0.7813
39	0.6807	0.6293	0.7771	0.8098
40	0.6981	0.6428	0.7660	0.8391
41	0.7156	0.6561	0.7547	0.8693
42	0.7330	0.6691	0.7431	0.9004
43	0.7505	0.6820	0.7314	0.9325
44	0.7679	0.6947	0.7193	0.9657

TABLE 9-1 Table of Trigonometric Functions (cont)

Angle	Radians	Sine	Cosine	Tangent
45	0.7854	0.7071	0.7071	1.0000
46	0.8029	0.7193	0.6947	1.0355
47	0.8203	0.7314	0.6820	1.0724
48	0.8378	0.7431	0.6691	1.1106
49	0.8552	0.7547	0.6561	1.1504
50	0.8727	0.7660	0.6428	1.1918
51	0.8901	0.7771	0.6293	1.2349
52	0.9076	0.7880	0.6157	1.2799
53	0.9250	0.7986	0.6018	1.3270
54	0.9425	0.8090	0.5878	1.3764
55	0.9599	0.8192	0.5736	1.4281
56	0.9774	0.8290	0.5592	1.4826
57	0.9948	0.8387	0.5446	1.5399
58	1.0123	0.8480	0.5299	1.6003
59	1.0297	0.8572	0.5150	1.6643
60	1.0472	0.8660	0.5000	1.7321
61	1.0647	0.8746	0.4848	1.8040
62	1.0821	0.8829	0.4695	1.8807
63	1.0996	0.8910	0.4540	1.9626
64	1.1170	0.8988	0.4384	2.0503
65	1.1345	0.9063	0.4226	2.1445
66	1.1519	0.9135	0.4067	2.2460
67	1.1694	0.9205	0.3907	2.3559
68	1.1868	0.9272	0.3746	2.4751
69	1.2043	0.9336	0.3584	2.6051
70	1.2217	0.9397	0.3420	2.7475
71	1.2392	0.9455	0.3256	2.9042
72	1.2566	0.9511	0.3090	3.0777
73	1.2741	0.9563	0.2924	3.2709
74	1.2915	0.9613	0.2756	3.4874
75	1.3090	0.9659	0.2588	3.7321
76	1.3265	0.9703	0.2419	4.0108
77	1.3439	0.9744	0.2250	4.3315
78	1.3614	0.9781	0.2079	4.7046
79	1.3788	0.9816	0.1908	5.1446
80	1.3963	0.9848	0.1736	5.6713
81	1.4137	0.9877	0.1564	6.3138
82	1.4312	0.9903	0.1392	7.1154
83	1.4486	0.9925	0.1219	8.1443
84	1.4661	0.9945	0.1045	9.5144
85	1.4835	0.9962	0.0872	11.4301
86	1.5010	0.9976	0.0698	14.3007
87	1.5184	0.9986	0.0523	19.0811
88	1.5359	0.9994	0.0349	28.6363
89	1.5533	0.9998	0.0175	57.2900

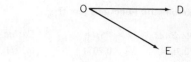

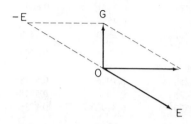

FIGURE 9-11   Vector subtraction by parallelogram.

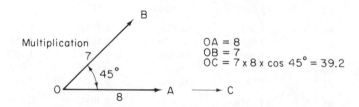

$OA = 8$
$OB = 7$
$OC = 7 \times 8 \times \cos 45° = 39.2$

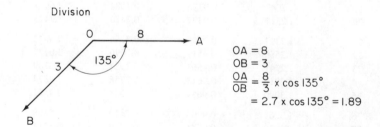

$OA = 8$
$OB = 3$
$\dfrac{OA}{OB} = \dfrac{8}{3} \times \cos 135°$
$= 2.7 \times \cos 135° = 1.89$

FIGURE 9-12   Vector multiplication and division.

9-8.4 Practical vector multiplication and division

Vectors can also be used for multiplication and division of out-of-phase values. Both of these functions require the use of trigonometry, as shown in Fig. 9-12.

9-9 INDUCTANCE IN CIRCUITS

As in the case of resistors, inductors (coils) can be connected in series, parallel, and in series-parallel. Where no interaction of magnetic fields is produced by the inductance, the equations are the same as for resistors, as shown in Fig. 9-13.

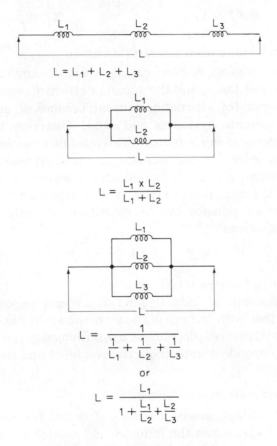

$$L = L_1 + L_2 + L_3$$

$$L = \frac{L_1 \times L_2}{L_1 + L_2}$$

$$L = \frac{1}{\dfrac{1}{L_1} + \dfrac{1}{L_2} + \dfrac{1}{L_3}}$$

or

$$L = \frac{L_1}{1 + \dfrac{L_1}{L_2} + \dfrac{L_2}{L_3}}$$

FIGURE 9-13   Inductance in circuits.

9-9.1 Inductive reactance and impedance of inductors

When current flows through an inductor (or coil), a magnetic field is set up around the inductance. If the current is alternating, the magnetic field will vary (build up and collapse). The buildup and collapse of the magnetic field will, in itself, produce another current in the inductor. This self-induced current is in opposition to the initial or external current. This opposition is known as inductive reactance and is expressed in ohms.

If an inductor has no dc resistance, its inductive reactance could be substituted for resistance in the conventional Ohm's law equations:

$$E = I \times X_L \qquad I = \frac{E}{X_L} \qquad X_L = \frac{E}{I}$$

where X_L is the inductive reactance (in ohms).

This is not possible in practical applications. Any inductor must have some dc resistance, and the actual or effective resistance will be somewhat higher for alternating current because of magnetic skin effect, eddy currents, hysteresis, and so on. Therefore, there is some *effective resistance* in any inductor. This resistance has the same effect as a resistor in series with the inductor, and is so represented.

The combined effect of the inductance and effective resistance is known as the *impedance*, which is also expressed in ohms. The impedance of an inductor can be substituted directly in the basic Ohm's law equations:

$$E = I \times Z \qquad I = \frac{E}{Z} \qquad Z = \frac{E}{I}$$

where Z is the impedance (in Ω).

The relationship of inductive reactance and impedance of inductors, together with the calculations, is shown in Fig. 9-14. Note that inductive reactance is dependent upon frequency and inductance; impedance is dependent upon inductive reactance and resistance.

9-9.2 Ac voltages in inductive circuits

Unlike dc voltages across resistors, it is not possible simply to add the ac voltages across the inductor and resistor in a series circuit to obtain the source voltage. Instead, the individual voltages must be added vectorially. Figure 9-15 shows the relationship of voltages together with the calculations for voltages in series RL circuits. The diagram and equation show that the voltage across the resistance E_R is equal to the current times the resistance, whereas the voltage across the inductance is equal to the same current times the inductive reactance. The source voltage E_{source} is equal to the vector sum of the two voltages.

Effective resistance (R)
(ohms)

Inductive reactance (X_L)
(ohms)

Z

Series
impedance

$$Z = \sqrt{R^2 + X_L^2}$$

Effective resistance (R)
(ohms)

(X_L)

Inductive
reactance
(ohms)

Parallel
impedance

$$Z = \frac{R \times L}{\sqrt{R^2 + X_L^2}}$$

Inductive reactance

$X_L = 6.28 \times$ frequency (in Hz) $\times$ inductance (in Henrys)

$$L = \frac{X_L}{6.28\ F} \qquad\qquad F = \frac{X_L}{6.28\ L}$$

FIGURE 9-14   Calculations for inductive reactance and impedance in resistance-inductance circuits.

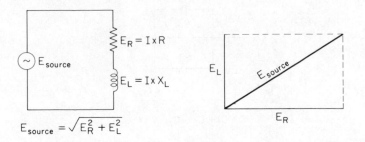

$E_R = I \times R$

$E_L = I \times X_L$

$\sim$ E_{source}

E_L E_{source}

E_R

$$E_{source} = \sqrt{E_R^2 + E_L^2}$$

FIGURE 9-15   Adding alternating current voltages in a series resistance-inductance circuit.

9-10 CAPACITANCE IN CIRCUITS

Capacitors function to store an electrostatic charge. Direct current can be stored in a capacitor, but alternating current will pass round the capacitor. Capacitors can be used in a variety of circuit combinations. The majority of circuits are, however, a version of the basic series, parallel, or series–parallel arrangements shown in Fig. 9-16.

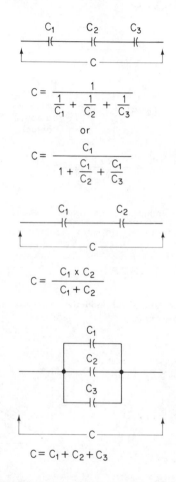

FIGURE 9-16   Capacitance in circuits.

The *farad* is the basic unit of capacitance, but since the farad is too large for practical applications, submultiples are used. The most common submultiples are the microfarad (mF, one millionth of a farad) and the picofarad (pF, one millionth of one millionth of a farad). The term "picofarad" has replaced the previously used term "micromicrofarad."

9-10.1 Capacitive reactance and impedance of capacitors

It is convenient to regard capacitive reactance as the opposition of a capacitor to alternating currents. Direct current will not pass through a capacitor. Therefore, the opposition is infinite. Pulsating current and alternating current will pass around a capacitor, but the process of alternately charging and discharging the capacitor produces some opposition. This opposition is termed *capacitive reactance* and is expressed in ohms.

If a capacitor is in a circuit where there is no dc resistance, the capacitive reactance can be substituted in the conventional Ohm's law equations

$$E = I \times X_C \qquad I = \frac{E}{X_C} \qquad X_C = \frac{E}{I}$$

where X_C is the capacitive reactance in ohms.

This is not possible in practical applications. Any circuit will have some dc resistance (even the capacitor terminals or conductors will present some resistance), and the actual or effective resistance will be somewhat higher for alternating current because of skin effect, eddy currents, hysteresis, and so on. Therefore, there is some effective resistance in any capacitive circuit. This resistance has the same effect as a resistor in series with the capacitor and is so represented.

The combined effect of the capacitance and effective resistance is known as the *impedance*, which is also expressed in ohms. The impedance of a capacitor can be substituted directly in the basic Ohm's law equations

$$E = I \times Z \qquad I = \frac{E}{Z} \qquad Z = \frac{E}{I}$$

where Z is the impedance in ohms.

The relationship of capacitive reactance and impedance of capacitors, together with the calculations, is shown in Fig. 9-17. Note that capacitive reactance is dependent upon frequency and capacitance; impedance is dependent upon capacitive reactance and resistance.

9-10.2 AC voltages in capacitive circuits

Unlike dc voltages across resistors, it is not possible simply to add the ac voltages across the capacitor and resistor in a series circuit to obtain the source voltage. Instead, the individual voltages must be

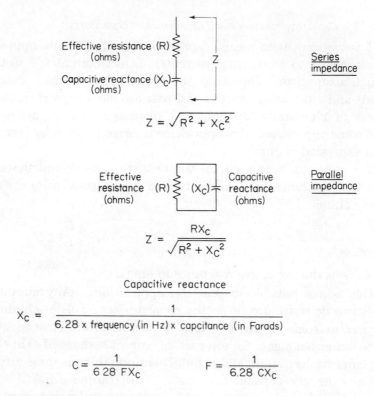

$$Z = \sqrt{R^2 + X_C^2}$$

$$Z = \frac{RX_C}{\sqrt{R^2 + X_C^2}}$$

Capacitive reactance

$$X_C = \frac{1}{6.28 \times \text{frequency (in Hz)} \times \text{capcitance (in Farads)}}$$

$$C = \frac{1}{6.28\, FX_C} \qquad F = \frac{1}{6.28\, CX_C}$$

FIGURE 9-17   Calculations for capacitive reactance and impedance in resistance-capacitance circuits.

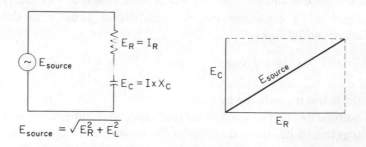

$$E_{source} = \sqrt{E_R^2 + E_L^2}$$

FIGURE 9-18   Aiding alternating current voltages in a series resistance-capacitance circuit.

added vectorially. Figure 9-18 shows the relationship of voltages together with the calculations for voltages in series RC circuit. The diagram and equation show that the voltage across the resistance E_R is equal to the current times the resistance, whereas the voltage across the capacitance is equal to the same current times the capacitive reactance. The source voltage E_{source} is equal to the vector sum of the two voltages.

9-11 PHASE ANGLE, IMPEDANCE, AND POWER FACTOR RELATIONSHIPS

In alternating currents, a pair of currents, a pair of voltages, or the voltage and current need not be in step. One current may lead the other, one voltage may lag the other, or the current may lag the voltage. The amount of lead or lag is termed *phase angle* and is expressed by the Greek letter theta, θ.

Since most alternating currents are sine waves, they have periodic waveforms where the lead or lag can be measured along the X axis. Out-of-phase voltages (or currents) can be added vectorially to produce a resultant voltage (or current).

In a pure resistive circuit there will be no lag or lead between the current and voltage (in phase); the phase angle is 0. In a pure inductive circuit, the current lags the voltage by $90°$; the phase angle is $-90°$. In a pure capacitive circuit, the current leads the voltage by $90°$; the phase angle is $+90°$. These relationships are shown in Fig. 9-19.

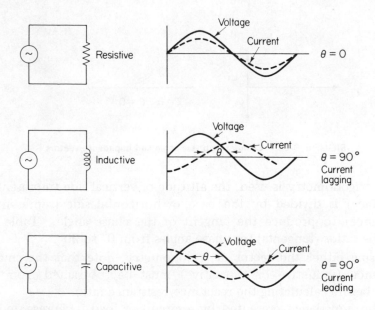

FIGURE 9-19   Relationship of phase angle in pure resistive, inductive, and capacitive circuits.

In practical applications most alternating current circuits are not pure inductive or capacitive but may be pure resistive. Often circuits are a combination of all three factors (usually resistive and inductive, such as motors, fluorescent lamps, and circuits involving transformers). In any case the *ratio of reactance to resistance* determines the phase angle. This may be calculated vectorially or by means of trigonometry. This relationship is shown in Fig. 9-20.

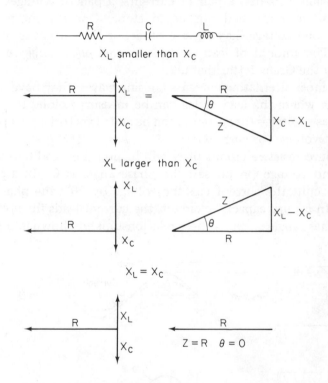

FIGURE 9-20   Calculating phase angle and impedance vectors.

If trigonometry is used, the altitude or vertical side (representing reactance) is divided by the base or horizontal side (representing resistance) to produce the tangent of the phase angle. Table 9-2 lists the ratios (tangents) for phase angles from $0°$ to $90°$.

Using either the vector or trigonometry, note that the smaller reactance (inductive or capacitive) must be subtracted from the larger before calculating the reactance/resistance ratio.

The impedance presented by a circuit or load is represented by

TABLE 9-2 Phase Angle Versus Ratio of Reactance Divided by Resistance

Phase Angle	Ratio	Phase Angle	Ratio	Phase Angle	Ratio
0	0.0000	30	0.5774	60	1.7321
1	0.0175	31	0.6009	61	1.8040
2	0.0349	32	0.6249	62	1.8870
3	0.0524	33	0.6494	63	1.9626
4	0.0699	34	0.6745	64	2.0503
5	0.0875	35	0.7002	65	2.1445
6	0.1051	36	0.7265	66	2.2460
7	0.1228	37	0.7536	67	2.3559
8	0.1405	38	0.7813	68	2.4751
9	0.1584	39	0.8098	69	2.6051
10	0.1763	40	0.8391	70	2.7475
11	0.1944	41	0.8693	71	2.9042
12	0.2126	42	0.9004	72	3.0777
13	0.2309	43	0.9325	73	3.2709
14	0.2493	44	0.9657	74	3.4874
15	0.2679	45	1.0000	75	3.7321
16	0.2867	46	1.0355	76	4.0108
17	0.3057	47	1.0724	77	4.3315
18	0.3249	48	1.1106	78	4.7046
19	0.3443	49	1.1504	79	5.1446
20	0.3640	50	1.1918	80	5.6713
21	0.3839	51	1.2349	81	6.3138
22	0.4040	52	1.2799	82	7.1154
23	0.4245	53	1.3270	83	8.1443
24	0.4452	54	1.3764	84	9.5144
25	0.4663	55	1.4281	85	11.4301
26	0.4877	56	1.4826	86	14.3007
27	0.5095	57	1.5399	87	19.0811
28	0.5317	58	1.6003	88	28.6363
29	0.5543	59	1.6643	89	57.2900

the hypotenuse of the triangle. Thus impedance is dependent upon resistance, reactance, and phase angle. The calculations for phase angle and impedance of common alternating current circuits are shown in Figs. 9-21 through 9-27.

9-11.1 Power factor relationships

The power factor of a circuit or load is also directly related to resistance, reactance, impedance, and phase angle, as shown in Figs. 9-21 through 9-27.

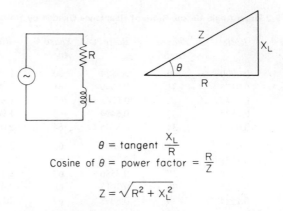

$$\theta = \text{tangent } \frac{X_L}{R}$$
$$\text{Cosine of } \theta = \text{power factor} = \frac{R}{Z}$$

$$Z = \sqrt{R^2 + X_L^2}$$

FIGURE 9-21   Phase angle and impedance of series RL circuit.

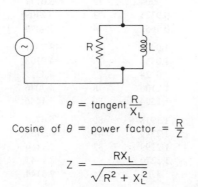

$$\theta = \text{tangent } \frac{R}{X_L}$$
$$\text{Cosine of } \theta = \text{power factor} = \frac{R}{Z}$$

$$Z = \frac{RX_L}{\sqrt{R^2 + X_L^2}}$$

FIGURE 9-22   Phase angle and impedance of parallel RL circuit.

In any reactive device (inductive or capacitive), current cannot be in phase with the voltage because of the reactance. The amount of angular displacement between voltage and current depends on the ratio or relative amount of resistance to reactance. Because the current and voltage do not reach maximum at the same instant, the real power of the circuit or load must be $E \times I$, multiplied by some factor less than 1. This factor compensates for the phase displacement between voltage and current. The factor is called the *power factor* of the load and is equal to the *cosine* of the angle between the voltage across and the current drawn by the load. The power factor is also equal to the ratio of the load resistance to load impedance. Thus

$$\text{power factor} = \text{cosine } \theta = \frac{R}{Z}$$
$$\text{power} = EI \text{ cosine } \theta = EI \times \text{power factor}$$

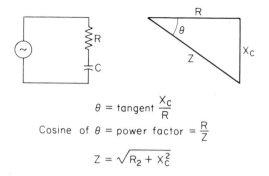

$$\theta = \text{tangent } \frac{X_C}{R}$$

$$\text{Cosine of } \theta = \text{power factor} = \frac{R}{Z}$$

$$Z = \sqrt{R_2 + X_C^2}$$

FIGURE 9-23   Phase angle and impedance of series RC circuit.

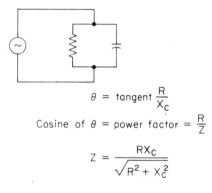

$$\theta = \text{tangent } \frac{R}{X_C}$$

$$\text{Cosine of } \theta = \text{power factor} = \frac{R}{Z}$$

$$Z = \frac{RX_C}{\sqrt{R^2 + X_C^2}}$$

FIGURE 9-24   Phase angle and impedance of parallel RC circuit.

Where X_L is larger than X_C

$$\theta = +90°$$
$$Z = X_L - X_C$$

Where X_C is larger than X_L

$$\theta = -90°$$
$$Z = X_C - X_L$$

FIGURE 9-25   Phase angle and impedance of series LC circuit.

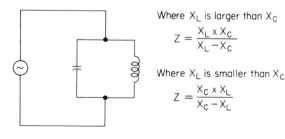

Where X_L is larger than X_C

$$Z = \frac{X_L \times X_C}{X_L - X_C}$$

Where X_L is smaller than X_C

$$Z = \frac{X_C \times X_L}{X_C - X_L}$$

When $X_L = X_C$
Phase angle is zero and Z is infinite

FIGURE 9-26   Phase angle and impedance of parallel LC circuit.

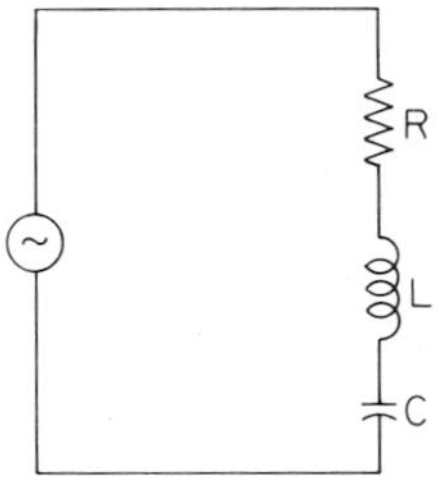

Cosine of θ = power factor

$$= \frac{R}{Z}$$

Where X_L is larger than X_C

$$Z = \sqrt{R^2 + (X_L - X_C)^2}$$

$$\theta = \text{tangent } \frac{X_L - X_C}{R}$$

Where X_C is larger than X_L

$$Z = \sqrt{R^2 + (X_C - X_L)^2}$$

$$\theta = \text{tangent } \frac{X_C - X_L}{R}$$

FIGURE 9-27   Phase angle and impedance of series RLC circuit.

9-11.2 Finding impedances in series and parallel circuits

When impedances or reactances are in series, they must be added, mathematically or by vectors. In a series circuit the current through each component is the same. Therefore, if the current is known and the individual reactances are known, the voltage across each component can be calculated using

$$E = IE \qquad E = IX_C \qquad or \qquad E = IX_L$$

If the current is not known, a current can be assumed to produce theoretical voltages.

Either way, the voltages can be added by vector calculation to find the total voltage, using

$$E_T = \sqrt{E_R^2 + (E_{XL} - E_{XC})^2} \qquad or \qquad E_T = \sqrt{E_R^2 + (E_{XC} - E_{XL})^2}$$

With the total voltage calculated, the total impedance of the network can be calculated using

$$Z = \frac{E}{I}$$

Going further, if the reactance or resistance is not known but the current and voltages are known, the individual reactances or resistance can be calculated using

$$R = \frac{E}{I} \qquad \text{or} \qquad X = \frac{E}{I}$$

If neither the voltage nor the reactance is known, the reactance can be calculated using the equations of Figs. 9-14 and 9-17. This, however, requires that frequency and capacitance or inductance be known.

The steps for solving series impedance problems are shown in Fig. 9-28.

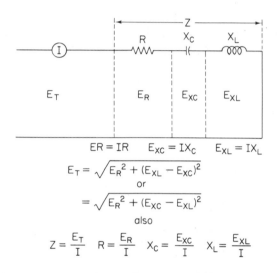

FIGURE 9-28   Finding impedance in series LRC circuits.

When impedances or reactances are in parallel, their currents must be added, mathematically or by vectors. In a parallel circuit the voltage across each component is the same. Therefore, if the voltage is known and the individual reactances are known, the current through each component can be calculated using

$$I = \frac{E}{R} \qquad I = \frac{E}{X_C} \qquad \text{or} \qquad I = \frac{E}{X_L}$$

If the voltage is not known, a voltage can be assumed to produce theoretical currents. Either way, the currents can be added by vector calculation to find the total current, using

$$I_T = \sqrt{I_R^2 + (I_{XL} - I_{XC})^2} \qquad \text{or} \qquad I_T = \sqrt{I_R^2 + (I_{XC} - I_{XL})^2}$$

With the total current calculated, the total impedance of the network can be calculated using

$$Z = \frac{E}{I}$$

Going further, if the reactance or resistance is not known but the voltage and currents are known, the individual reactances or resistances can be calculated using

$$R = \frac{E}{I} \qquad \text{or} \qquad X = \frac{E}{I}$$

If neither the current nor the reactance is known, the reactance can be calculated using the equations of Figs. 9-14 and 9-17. This requires, however, that frequency and capacitance or inductance be known.

The steps for solving parallel impedance problems are shown in Fig. 9-29.

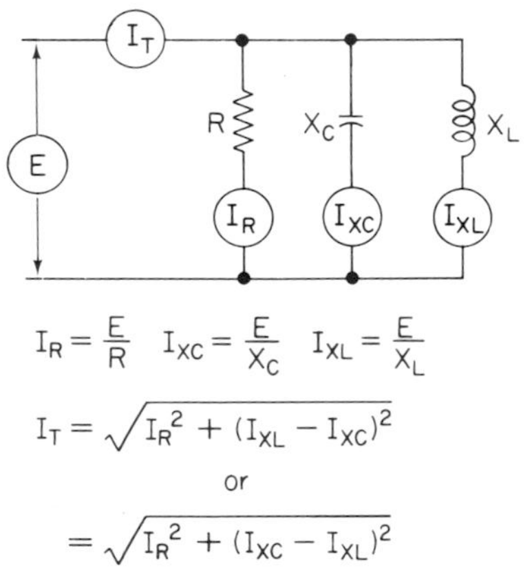

$$I_R = \frac{E}{R} \quad I_{XC} = \frac{E}{X_C} \quad I_{XL} = \frac{E}{X_L}$$

$$I_T = \sqrt{I_R^2 + (I_{XL} - I_{XC})^2}$$

or

$$= \sqrt{I_R^2 + (I_{XC} - I_{XL})^2}$$

also

$$Z = \frac{E}{I_T} \quad R = \frac{E}{I_R} \quad X_C = \frac{E}{I_{XC}} \quad X_L = \frac{E}{I_{XL}}$$

FIGURE 9-29   Finding impedance in parallel LRC circuits.

9-12 POWER TRIANGLE

A typical electrical distribution system consists of several different loads, all connected in parallel. Some of the loads are pure resistive, such as heating elements and incandescent lamps. Other loads can be part resistive and part inductive, such as electric motors and fluorescent lamps. A typical load is not likely to be pure capacitive unless it is a special load consisting of capacitors (or a synchronous motor) connected to the correct power factor, as discussed in Section 9-13.

Unless all the loads are pure resistive, it is not practical to evaluate the total load (or even one of the loads) on the basis of resistance alone. If the total load or any part of the load is inductive, then inductive reactance must be included. Since resistance and reactance produce an impedance, the total impedance factor must be considered. And, since impedance is involved, power factor must also be considered.

It is possible to combine all these factors for a single load in an electrical system, or several loads, by means of a *power triangle* as shown in Fig. 9-30. Such a triangle is a graphic means of showing load relationships. As shown, the power triangle is made up by multiplying the basic impedance triangle by the current through the load. In the example of Fig. 9-30 the load is part resistive and part reactance (inductive) in series. Thus the current is the same in both the reactance and the resistance. When these factors are multiplied by the current, the voltage is found and the result is a *voltage triangle*. (This is produced by the basic Ohm's law relationship $E = IR$.)

When all the factors in the voltage triangle are again multiplied by current, a power triangle is formed. (This is produced by the basic Ohm's law relationship $P = EI$.) The power triangle has the same shape as the original impedance triangle (same angles and same ratios). However, the horizontal side of the power triangle is expressed in watts (instead of resistance), and the hypotenuse is expressed in volt-amperes (VA) (instead of impedance). Another way of expressing this is that the wattage of a power triangle has the same relationship to VA as resistance does to impedance. Or, resistance divided by impedance produces the same fraction or ratio as occurs when wattage is divided by VA. In either case the result is power factor and is equal to the cosine of the phase angle.

Note that the volt-amperes unit is sometimes called *apparent power* since it is the power that appears to be used ($P = EI$). However, true power is found when the VA is multiplied by the power factor. For this reason it is possible to measure both the voltage and current of the load (with a voltmeter and an ammeter) and the watt-

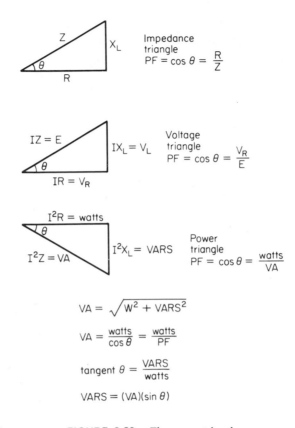

FIGURE 9-30   The power triangle.

age of the same load (with a wattmeter) and find that the two do not agree.

The vertical side of the power triangle represents *reactive power*, which, by itself, is an imaginary quantity. For this reason reactive power is expressed in an imaginary electrical unit of *vars* (voltamperes reactive) or *kvars* (1000 voltamperes reactive).

Most power triangles are drawn with the vertical side going down, since most practical electrical loads are inductive, causing the current to lag the voltage. If the load is capacitive, the current leads the voltage, and the vertical side of the triangle is drawn going up. Note that vars is equal to VA × the sine of the phase angle. For this reason the sine of the phase angle is often called the *reactive factor* (just as the cosine of the phase angle is called the power factor).

9-12.1 Example of power triangle

Figure 9-31 shows how the power triangle and the related equations can be used to solve a practical problem.

Suppose that a motor draws 8 A from a 120-V line. The motor is rated at 720 W (just below 1 hp). Find the VA, the power factor, the phase angle, and the vars.

VA: Multiply voltage (120) by current (8) to find a VA of 960 VA.

Power factor: Divide the wattage (720) by the VA (960) to find a power factor of 0.75.

Phase angle: The cosine of 0.75 is 41.4°, as shown in Table 9-1.

Vars: Multiply VA (960) by the sine of 41.4° (Table 9-1 shows a sine of 0.6613), or 960 × 0.6613 – 635 vars.

A similar problem can be stated in another way. Assume that a motor is rated as 10 hp and operates on 120 V with a power factor of 0.8. What current will be drawn?

First find the wattage. One horsepower = 746 W; 10 hp = 7460 W.

Divide the wattage (7460) by the power factor 0.8 to find a VA of 9325.

Divide the VA (9325) by the voltage (120) to find a current of 77.7 A.

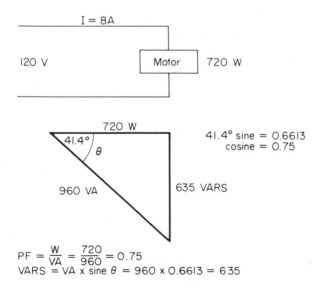

FIGURE **9-31**   Example of power triangle calculations.

9-13 ADDING LOADS WITH THE POWER TRIANGLE

If all the loads in a particular distribution system have the same power factor, it is relatively easy to find the total current of the system. Simply add all the rated wattages, divide by the power factor to find VA, then divide by the system voltage to find the current. However, if the loads have different power factors, it is necessary to use vectors to add the wattage. This can be done with the power triangle, as shown in Fig. 9-32.

   Assume that there are three 30-kW loads connected in parallel

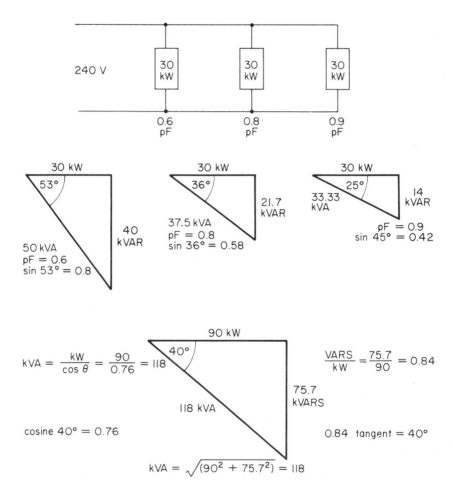

FIGURE 9-32   Adding loads of different power factors with the power triangle.

across a 240-V system. The loads have power factors of 0.6, 0.8, and 0.9. Find the total current of the system as well as the overall power factor for the system.

The 0.6 power factor load has a phase angle of approximately 53° (Table 9-1) and a kVA of 50 (30 kW/0.6). The sine of 53° is approximately 0.8. Thus the load has 40 kvars (50 kVA × 0.8).

The 0.8 power factor load has a phase angle of approximately 36° (Table 9-1) and a kVA of 37.5 (30 kW/0.8). The sine of 36° is approximately 0.58. Thus the load has 21.7 kvars (37.5 kVA × 0.58).

The 0.9 power factor load has a phase angle of approximately 25° (Table 9-1) and a kVA of 33.33 (30 kW/0.9). The sine of 25° is approximately 0.42. Thus the load has 14 kvars (33.33 kVA × 0.42).

The total kW is 90 (30 + 30 + 30), and the total kvars is 75.7 (40 + 21.7 + 14). The total kvars/total kW ratio is 75.7/90, or approximately 0.84. The tangent 0.84 is for an angle of approximately 40° The cosine of 40° is approximately 0.76. Thus the overall power factor is 0.76.

With an overall power factor of 0.76 and a total wattage of 90 kW, the total kVA is approximately 118 (90 kW/0.76). This can be confirmed by the equation $VA = \sqrt{W^2 + vars^2}$. With a kVA of 118 and a system voltage of 240 V, the total current is approximately 493 A.

9-14 CORRECTING THE POWER FACTOR

An increase in power factor for an electrical system results in lower current, assuming a given voltage and a given wattage. Generally, the voltage and wattage of a power distribution system are specified or are determined by the number of loads involved. Either way, the voltage and wattage cannot be changed. The only way to increase efficiency of the system is to increase the power factor. Many utility companies require a minimum power factor, typically 0.8 to 0.85 or better. In some cases the utility companies assess an extra charge if the power factor for a system is below 0.8. This depends upon local regulations. Most utility companies are not interested in the power factors of individual loads but in the overall system power factor, as measured at the service entrance. (This is determined when the service entrance wattmeter reading is divided by the VA at the service entrance.)

No matter what regulations or conditions exist, it is always good design practice to keep the power factor as high as practical. This can be done in two ways: by adding capacitors across the line or by adding a synchronous motor across the line. Both methods are discussed in this section. No matter which method is used, the basic principle is the same.

Most loads (that are not pure resistive) are inductive (typically electric motors and fluorescent lamps). Inductive loads have inductive reactance. If a capacitor is added across the line, the capacitor produces some capacitive reactance, even though the capacitor draws no current. The capacitive reactance cancels the inductive reactance, thus reducing the overall resistance/reactance ratio and increasing the power factor. A synchronous motor also produces a capacitive effect on the line. Both devices cause the current to lead the voltage. This lead offsets any lag produced by the inductive load, reducing the phase angle to zero (or near zero), and increasing the power factor to 1 (or near 1).

9-14.1 Power factor correction with capacitors

Assume that a system load is measured at the service entrance. The load is rated at 60 kW (six 10-kW electric motors). However, the current is 333.33 A at 240 V. This shows an apparent power of 80 kVA (240 V × 333.33 A) and a power factor of 0.75 (60 kW/80 kVA). Find what value capacitor is required to raise the power factor to 0.85, so the current is reduced to 294 A (60 kW/0.85 = 70.588 kVA; 70.588 kVA/240 V = 294 A).

The angle with a cosine of 0.85 is approximately 32° (Table 9-1). The sine of 32° is approximately 0.53. If the desired kVA is 70.588 (as it will be if the power factor is 0.85), the required kvars is 0.53 × 70.588, or approximately 37.4 kvars.

The kvars of the existing system (without correction) is equal to the existing VA (80 kVA) times the sine of the existing phase angle. The angle with a cosine of 0.75 (existing) is approximately 41° (Table 9-1). The sine of 41° is approximately 0.66. Thus the existing (lagging) kvars is 0.66 × 80 kVA, or approximately 52.8 kvars.

With an existing 52.8 kvars and a desired 37.4 kvars, the capacitors must supply 15.4 kvars (leading).

Capacitors used for power factor correction are rated as to kvars as well as voltage. However, it may be necessary to find the capacitance value that will produce 15.4 kvars in some cases. The following steps describe the procedure.

Find the leading reactive current (imaginary) that occurs with 15.4 kvars and 240 V, 15.4 kvars/240 V = 64.25A (reactive).

With a reactive current of 64.25 A and a voltage of 240 V, the capacitive reactance is approximately 0.267 Ω ($X_C = I_C/E$ = 64.24/ 240 = 0.267).

The capacitance required to produce a reactance of 0.267 Ω at 60 Hz (the usual line frequency) is approximately 10,000 μF. This is a large capacitance value and would probably be made up of many smaller capacitors connected in parallel (such as one hundred 100-μF capacitors). The capacitors must have a voltage rating of at least 240 V and preferably 1.5 times the line voltage, or 240 V × 1.5 = 360 V.

9-14.2 Power factor correction with synchronous motors

When a synchronous motor is connected in parallel with an electrical distribution system and the field current is properly adjusted in relation to the load, the current drawn by the motor will lead the voltage. As far as reactance is concerned, the synchronous motor appears as a capacitor in parallel with the line. Thus a large synchronous motor driving a fixed load (ventilating fan, etc.) can raise the power factor of the entire system. It is not even necessary for the motor to drive a load, although such a design is usually wasteful, since the motor will draw some current. When a synchronous motor is used for power factor correction, the motor is sometimes called a *synchronous condenser* or *synchronous capacitor.*

Assume that a synchronous motor (without a load) is used to correct the power factor of the system described in Section 9-13.1. With such an arrangement, the field current of the motor must be adjusted to draw 64.25 A (reactive). In theory the motor produces no power. In practice the motor will produce some power, even if the drive shaft is spinning in free air. This power is added to the overall load and thus increase the current. However, since the effect of adding leading kvars to the system produces a drastic reduction in current, the load produced by the motor can be neglected in practical design.

Now assume that a synchronous motor (with a 5-kW) load is used to correct the power factor of the system described in Section 9-13.1. The original load is 60 kW, 80 kVA, power factor 0.75, 240 V, 333.33 A, phase angle 41°, and 52.8 kvars.

With the motor added, the load is increased to 65 kW. However, the power factor is increased to 0.85, with a phase angle of $32°$. The desired kVA (with a power factor of 0.85) is 76.47. The sine of $32°$ is approximately 0.53.   If the desired kVA is 76.47, the required kvars is 0.53×76.47, or approximately 40.5.

With an existing (before correction) 52.8 kvars and a required 40.5 kvars, the synchronous motor must supply 12.3 kvars (leading).

The field current of the motor must be adjusted so that the kVA value includes both the 12.3 kvars and the 5-kW load.   As shown in Fig. 9-30, VA = $\sqrt{W^2 + \text{vars}^2}$.   Thus the motor must be adjusted for a kVA of $\sqrt{5^2 + 12.3^2}$ = approximately 13.3.   At 240 V the current is approximately 55 A for the synchronous motor.

Under these conditions the overall system current is now 76.47 kVA/240 V, or approximately 315 A.   Note that this is about 21 A higher than the 294 A required by the capacitor power factor correction described in Section 9-13.1.   However, the synchronous motor drives an additional 5-kW load, which requires the additional 21 A.

9-15 THREE-WIRE POWER DISTRIBUTION SYSTEM

The three-wire power distribution system shown in Fig. 9-33 is used in a great majority of the electrical systems. The system has the advantage of providing both 120 V and 240 V from the same wiring. The 240 V is used for large loads (such as large appliances) to reduce current. All other factors being equal, the current is reduced to one-half when 240 V is used instead of 120 V. The 120 V is used for lighting and small loads.

The utility company will often provide the three wires to the service entrance, but only two of the wires will be carried throughout the interior wiring. This provides 120 V for the interior and is used when the customer has no large loads. Should the customer add large loads (typically large appliances such as electric ranges, large electric clothes dryers, etc.), the third wire can be carried through the interior from the service entrance.

When only two wires are used in the interior, one wire must be the *neutral wire.   This neutral wire must be grounded (at the service entrance, Chapter 4), must never be fused, and must never be disconnected.*

The system shown in Fig. 9-33 is generally referred to as the Edi-

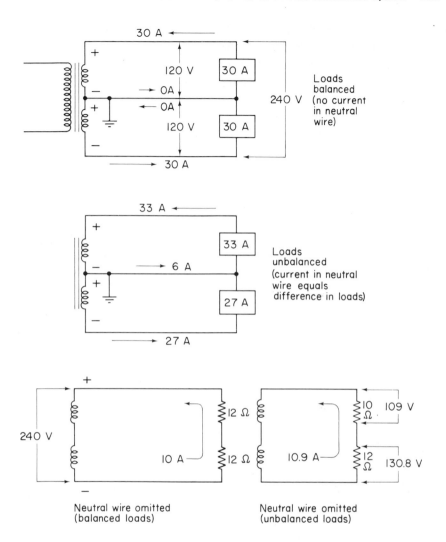

FIGURE 9-33   3-wire Edison system load arrangements.

son three-wire system.   The system is formed by connecting two 120-V power sources in series in such a way that their polarities cause the voltages to be additive.   This is shown by the polarity markings in Fig. 9-33. Of course, since alternating current is involved, the polarity markings are instantaneous.   However, at any given instant, the bottom 120-V supply is added to the top 120-V supply so that the total across the two supplies is 240 V.   There are several ways of producing such a supply.   The most common way is with a center-tapped transformer, as shown.   The secondary winding is actually two 120-V windings in series.

Ideally, the load should be distributed evenly between the two 120-V supplies.   That is, the loads should be the same on either side of the neutral wire.   For example, if the total load is 60 A, the branch circuits should be arranged so that 30 A is supplied by each of the 120-V sources.   Under these conditions the system is balanced, and there is no current flowing in the neutral wire.   That is, the current through the top load opposes the current through the bottom load in the neutral wire.   Since both currents are equal, they cancel and produce zero current in the neutral wire.

No matter how well balanced the system is designed, there is no guarantee that all the loads will be used simultaneously.   There is always the possibility of unbalance.   When this occurs, current will flow in the neutral wire.   The amount of current is the *difference* between the two 120-V load currents.   For example, if the load connected to the top 120-V source draws 33 A, and the bottom load draws 27 A (still a total current of 60 A), there is 6 A flowing in the neutral wire (33 A - 27 A = 6 A).   From a design standpoint, note that the neutral wire always carries less current than either of the other wires.   Thus the neutral wire size need never be larger than the other wires.   In practice, always use the same wire size for all three wires.

Never consider any design where the neutral wire is omitted. Even if all the loads require only 240 V, the neutral wire must be brought in and grounded at the service entrance.   It is sometimes assumed that if the 120-V loads are balanced, the neutral wire can be omitted, since it does not carry any current, and the voltage drop across each of the loads is equal.   Thus a 240-V source will divide, with 120 V across each load.   This may work, in theory but not in practice.   If there is the slightest unbalance, the voltages will not divide evenly, resulting in one voltage being too high, and one too low.   For example, assume that each of the loads are supposed to be 12 Ω, pure resistive.   The total series resistance of the loads is 24 Ω. With 24 Ω and 240 V, the current is 10 A.   The 10 A through both loads produces 120 V across each load.   Now assume that one load is only 10 Ω with the other remaining at 12 Ω.   The total load resistance is 22 Ω and the total current is 10.9 (240 V/22 Ω).   The voltage across the 10-Ω load is 109 V (10 $\times$ 10.9); the voltage across the

12-Ω load is 130.8 V. The high voltage could cause damage to the equipment. The low voltage will probably result in poor performance.

9-16 THREE-PHASE POWER DISTRIBUTION SYSTEM

A three-phase generator has the armature coils divided into three sections. The armature coils, regardless of their number, are grouped into sets so arranged that the induced voltage for one set of coils differs from the others by a third of a cycle (120°), as shown in Fig. 9-34. Three-phase generators are also known as *polyphase* generators, since they produce more than one phase of alternating current.

Three-phase systems (generators, transformers, motors, etc.) are classified according to connection into two groups: the *delta* (Δ) connection (so called since it resembles the Greek letter positioned sideways) and the Y (wye) or *star* connection. Sometimes the systems are mixed, as when a delta motor is powered by a wye generator through a wye-to-delta transformer.

The main advantage of three-phase systems is that increased power can be obtained without a corresponding increase in individual voltages and currents. Thus conductor sizes can be kept smaller. In a wye system this is possible since the line voltage is 1.73 ($\sqrt{3}$) times the voltage across each phase. In a delta system the power increase is possible since the line current is 1.73 times the current produced in each phase.

The terms *phase current* (I_{phase}) and *phase voltage* (E_{phase}) refer to currents and voltages in one phase or branch of a three-phase load.

The terms *line current* (I_{line}) and *line voltage* (E_{line}) refer to currents and voltages in one of the lines or connecting wires (conductors) between source and load.

Line current is the current that will be found if an ammeter is connected in any one line (conductor). In the case of a wye system, the line current and phase current have the same value.

Line voltage is the voltage that is found across any two lines (except the neutral conductor in wye). In the case of a delta system the line voltage and phase voltage have the same value.

The equations of Fig. 9-34 show the factors involved to find power in both delta and wye systems.

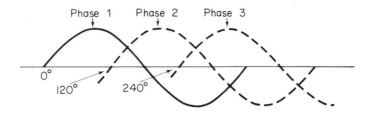

Phase 1   Phase 2   Phase 3

0°
120°
240°

θ = Phase angle between voltage and current in any one phase

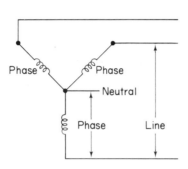

Phase   Phase

Neutral

Phase   Line

Wye, Star, or Y connections

$$E_{Phase} = \frac{E_{Line}}{1.73}$$

$$I_{Line} = I_{Phase}$$

$$E_{Line} = E_{Phase} \times 1.73$$

Total power =

$$E_{Line} \times 1.73 \times I_{Line} \times \cos \theta$$

or

$$3\ E_{Phase} \times I_{Phase} \times \cos \theta$$

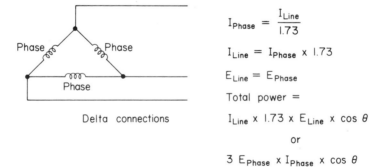

Phase   Phase

Phase

Delta connections

$$I_{Phase} = \frac{I_{Line}}{1.73}$$

$$I_{Line} = I_{Phase} \times 1.73$$

$$E_{Line} = E_{Phase}$$

Total power =

$$I_{Line} \times 1.73 \times E_{Line} \times \cos \theta$$

or

$$3\ E_{Phase} \times I_{Phase} \times \cos \theta$$

FIGURE 9-34   Basic connections and calculations for 3-phase systems (balanced loads).

9-16.1 Three-phase four-wire systems

In certain power systems, where both three-phase and single-phase are required, or where there is a possibility of unbalance in the loads, a neutral wire is used, as shown in Fig. 9-35. The neutral wire is grounded (at the service entrance when the utility company supplies the three-phase four-wire service) and serves two purposes.

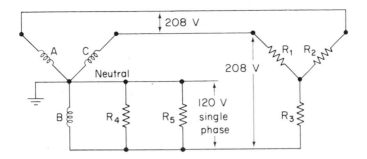

FIGURE 9-35   3-phase, 4-wire system, supplying both single-phase (120V) and 3-phase (280V).

First, the neutral wire permits single-phase to be obtained from a three-phase system. For example, each phase (A, B, and C) shown in Fig. 9-35 has a 120-V drop across it; this voltage is obtained with respect to the neutral wire and any one of the other wires. For the system shown, the voltage drop across B is used to supply 120-V single-phase to load circuits R_4 and R_5. These could be lights, heaters, or other devices requiring only 120-V single-phase. At the same time, a balanced load (R_1, R_2, and R_3) is furnished by the three-phase.

The neutral wire also serves the purpose of carrying the *difference current* when the loads are not balanced. When all three loads in a three-phase system are balanced (all three loads equal), there is no current in the neutral wire. However, if one phase has more (or less) current than the other phases, there will be current in the neutral wire (as discussed in Section 9-14).

In the case of the system in Fig. 9-35, the three-phase loads (R_1, R_2, and R_3) are balanced and there is no current in the neutral wire that results from these loads. However, phase B has the additional single-phase loads. Thus the neutral wire must be capable of carrying the current produced by R_4 and R_5.

9-16.2 Power in unbalanced three-phase loads

The equations shown in Fig. 9-34 are based on balanced three-phase loads. When the loads are not balanced, the calculations to find power and current are more difficult and involve the use of vectors.

Unbalanced delta. Figure 9-36 shows the equations and vectors necessary to find the power and line current in unbalanced delta systems. To find total power, the powers in each phase (phase power)

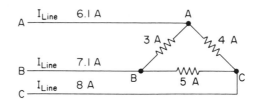

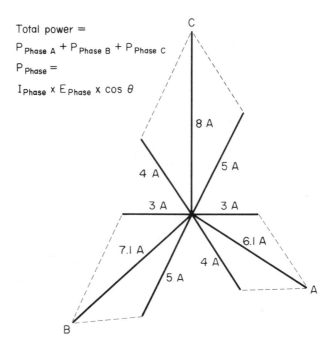

FIGURE 9-36   Power and current in unbalanced delta system.

must be added. In the example shown the phase currents are 3, 4 and 5 A and the phase (or line) voltage is 120 V. Thus the phase powers are 360, 480 and 600 W, and the total power is 1440 W. If the problem had been stated in reverse, with the wattage and voltage given, the phase currents could have been found when the wattage is divided by voltage (assuming a pure resistive load and a power factor of 1).

Design of a system such as shown in Fig. 9-36 requires that the line current be known, not the phase current. The conductors must carry line current; wire size, voltage drop, and so on, must be based on line current.

The current in any one line is the vector sum of the corresponding two-phase currents. For example, the current in line A is the vector sum of the 3-A and 4-A loads; line B is the 3-A and 5-A vector sum; line C is the 4-A and 5-A vector sum. These are relatively simple vectors to find, using the graphic method, when it is found that the angle between any of the *two-phase currents* is 60°.

As shown in Fig. 9-36, the current in line A is 6.1 A, the current in line B is 7.1 A, and the current in line C is 8 A.

Unbalanced wye. When there is any possibility of unbalance in a wye system, the neutral wire must be used. Thus the voltage across each load is the same, even though the currents may be different. The current in each line is equal to the current in the corresponding phase. If the wattage is given, the current (phase and line) can be found when the wattage is divided by the voltage.

If the wye system is unbalanced, there will be some current in the neutral wire. This current is the vector sum of the three line (or phase) currents. The angle between any two of the currents is 120°, as shown in Fig. 9-37.

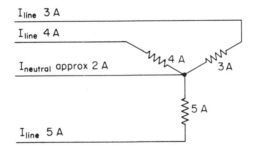

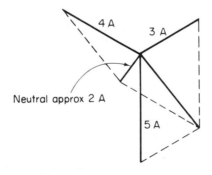

FIGURE 9-37   Current in unbalanced wye system.

Assume that the values for the system of Fig. 9-37 are the same as for the delta system (3, 4, and 5 A). As shown, the neutral wire current is about 2 A. As a rule of thumb, the current in the neutral wire is less than any of the line currents. Thus wire size for the neutral conductor need be no larger than for the other line conductors in the system.

9-17 POWER MEASUREMENT FOR ELECTRICAL DISTRIBUTION SYSTEMS

The *apparent power* or volt-amperes of an electrical distribution system can be measured with a voltmeter and an ammeter. If the load is inductive (or has any reactance), the apparent power will not be the true power. The true power drawn by a load, or all the loads in the system, can be measured directly by means of a *wattmeter*. The basic connections for a wattmeter are shown in Fig. 9-38. These connections are for single-phase power, either pure resistive or inductive.

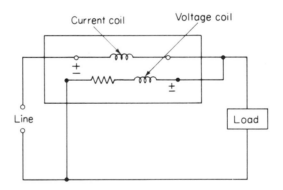

FIGURE 9-38   Basic wattmeter connections for single-phase power measurement.

A typical single-phase wattmeter has four terminals and two circuits. The *current coil* terminals (usually larger in size) are connected in series with the circuit, just as for an ammeter. The *potential coil* or *voltage coil* terminals are connected across the circuit, as in the case of a voltmeter. Note that the ± terminal of the current coil is connected to the *line* line of the circuit, and that the ± terminal of the voltage coil is connected to the *same* line lead as the current coil. Reversal of either winding will result in a backward deflection.

Wattmeters are designed to recognize power factor angle and to account for the power factor (if any) in their readings. The wattmeter multiplies the current passing through the current coil by the voltage impressed on the voltage coil, and then multiplies this product by the cosine of the angle between the two values. The result is the true wattage of the circuit being measured.

9-17.1 Three-phase power measurement

Several methods have been used for measurement of power in three-phase systems using wattmeters. The most common method in present use is shown in Fig. 9-39. This arrangement requires two wattmeters and applies to three-phase three-wire systems, balanced or unbalanced. The connections of Fig. 9-39 do not apply to three-phase four-wire systems (wye, with neutral wire), unless the wye system was perfectly balanced. Since the purpose of the neutral wire in a wye

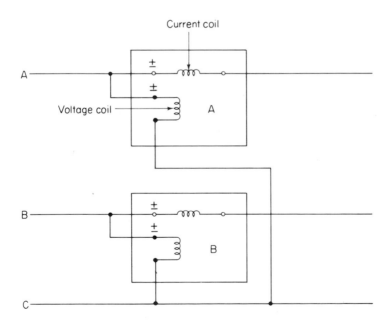

FIGURE 9-39   Wattmeter connections for 3-phase, 3-wire power measurement, balanced or unbalanced loads.

system is to handle possible unbalances in the loads, a special arrangement must be used for three-phase four-wire, as discussed in Section 9-16.2.

Note that in the two-wattmeter method of Fig. 9-39, all common connections are on the line side. Also note that the current of wattmeter A lags its voltage by $30°$, whereas the current of wattmeter B leads its voltage by $90°$. Should one wattmeter read back scale, it is only necessary to reverse the current connections.

If both wattmeters read the same value, this indicates that the load is balanced and that the power factor is one. The total wattage of the system is found by adding the two wattage readings (wattmeter A + wattmeter B = total system wattage).

If the load is balanced and each wattmeter reads a different value, this indicates that the power factor is not 1 (there is a combination of inductance and resistance in the loads). The total wattage of the system is still found by adding the two wattage readings. The power factor is found from the ratio of the two readings. The power factor can be found using the curve of Fig. 9-40 or by the equation on

$$\text{Tangent } \theta = 1.73 \ \frac{W_2 - W_1}{W_2 + W_1}$$

W_1 = Smaller reading

W_2 = Larger reading

Power factor = cosine θ

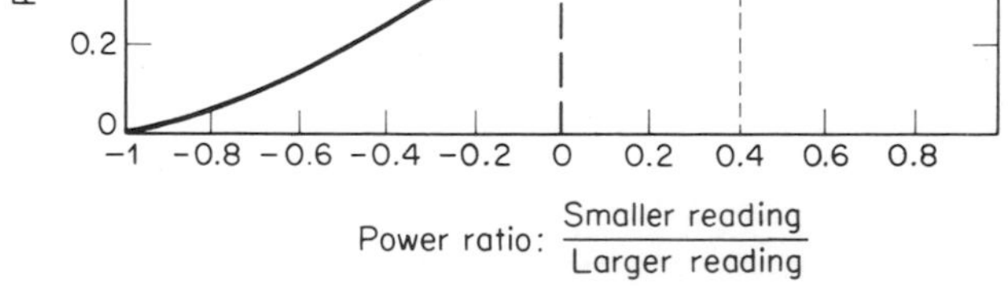

FIGURE 9-40   Relation between power factor and wattmeter ratio for balanced 3-phase loads.

Fig. 9-40. For example, assume that the wattmeter readings are 240 W and 600 W. Using the curve, the ratio is 0.4 (240/600) and the power factor is 0.8. Using the equation

$$1.73 \times \frac{600 - 240}{600 + 240} = \text{approximately } 0.74$$

the angle with a tangent of 0.74 is approximately 36.5° (Table 9-1). The cosine of 36.5° is approximately 0.8 (thus the power factor is 0.8).

In checking the curve of Fig. 9-40, note that the ratio is 1 at unity power factor (since the wattmeter readings are equal); the ratio is zero at 0.5 power factor (since the lower-reading wattmeter indicates zero); that the ratios are *negative* for all lower power factors; and that the ratio is −1.0 at zero power factor (since the wattmeter readings are again equal but the lower reading wattmeter gives reverse indication).

Keep in mind that the curve and equation of Fig. 9-40 are applicable only for *balanced* loads, when metered by the two-wattmeter method. Also, if one wattmeter reads zero, the power factor is 0.5. Should the power factor be less than 0.5, the angle between one wattmeter current and voltage will be greater than 90°. As a result that particular reading will be considered negative and must be subtracted from the other. As a practical matter, *the amount to subtract may be found* by reversing the current connections, so that the wattmeter indicates up-scale rather than in a reverse manner.

Hints on practical wattmeter measurements. With single-phase loads, a wattmeter will deflect backwards only if it is improperly connected. This is not true when the two-wattmeter method is used for three-phase, and the power factor is less than 0.5. When a wattmeter gives a reverse reading, it may be difficult to tell the difference between a low power factor or an improper connection. This can be avoided by paying careful attention to wattmeter polarity markings. The following procedure is recommended for connecting wattmeters for the two-wattmeter method.

1. Connect each wattmeter in the proper manner. That is, connect the ± terminal of the current coil on the line side of the line lead and the ± terminal of the voltage coil to the same line as its own current coil.
2. If either wattmeter deflection is backwards, it is an indication

of low power factor. Thus reverse the *current* coil connections and consider this reading a *negative* value.

9-17.2 Three-phase four-wire power measurement

The most common method of measuring three-phase four-wire systems (wye, with a neutral wire for unbalanced loads) is shown in Fig. 9-41. Note that all the voltage coils are connected to the neutral wire. This permits each phase to be measured by a separate wattmeter, and the total power is the sum of their readings (wattmeter A + wattmeter B + wattmeter C = total system wattage).

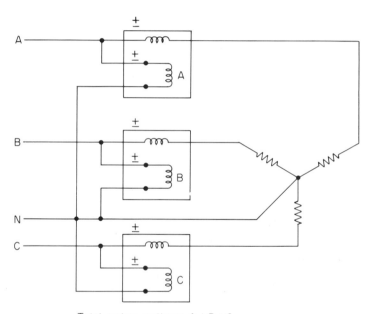

Total system wattage: A + B + C

FIGURE 9-41   Wattmeter connections for 3-phase, 4-wire power measurement, balanced or unbalanced loads.

9-18 ESTIMATING THE OPERATING COSTS OF ELECTRICAL APPLIANCES

Table 9-3 shows the annual operating cost of electrical appliances for normal use. The cost base is 2 cents/kWh. Of course, the costs will be different if the cost base is different or the appliance is not used the average number of hours.

TABLE 9-3 Annual Operating Cost of Electric Appliances for Normal Use
(*Courtesy* General Electric Company)

Appliance	Rated watts	Hours for 10 cents	Annual kWH	Cost	Appliance	Rated watts	Hours for 10 cents	Annual kWH	Cost
Air conditioner	1300	4	1,265	$25	Hot plate	1250	4	100	$2
(window)					Humidifier	70	70	140	3
Bed covering	170	30	140	3	Incinerator	605	8	665	13
Broiler	1375	4	95	2	Iron (hand)	1050	5	135	3
Clock	2	2500	17.5	0.35	Ironer (mangle)	1525	3	160	3
Clothes dryer	4800	1	960	19	Radio	80	62	90	2
Coffee maker	850	6	95	2	Radio-phonograph	105	48	105	2
Cooker (egg)	500	10	14	0.28	Range	11,720	1/2	1225	25
Deep fat fryer	1380	4	75	1.50	Refrigerator	235	21	460	9
Dehumidifier	240	20	380	8	Refrigerator–freezer	330	15	845	17
Dishwasher	1190	4	340	7	Refrigerator–freezer	425	12	1625	32
Electrostatic	60	83	265	5	(frostless)				
cleaner					Roaster	1345	4	205	4
Fan (attic)	375	13	310	6	Sewing machine	75	67	10	0.20
Fan (circulating)	85	60	40	1	Shaver	15	330	2	0.04
Fan (roll–about)	205	25	110	2	Sun lamp	290	17	15	0.30
Fan (window)	190	26	145	3	Television	255	20	345	7
Floor polisher	315	16	15	0.30	Television (color)	300	17	450	9
Food blender	290	17	15	0.30	Toaster	1110	5	35	1
Food freezer	300	17	915	18	Vacuum cleaner	540	9	40	1
Food mixer	125	40	10	0.20	Vibrator	40	125	2	0.04
Food waste	420	12	20	0.40	Waffle iron	1080	5	20	0.40
disposer					Washing machine	375	13	65	1
Fruit juicer	100	50	5	0.10	(automatic)				
Frying pan	1170	4	190	4	Washing machine	280	18	50	1
Germicidal lamp	20	250	135	3	(non-automatic)				
Grill (sandwich)	1050	5	30	0.60	Water heater	3000	2	4070	81
Hair dryer	300	17	7	0.14	(standard)				
Heat lamp	250	20	12	0.24	Water heater	4500	1	4475	98
(infared)					(quick recovery)				
Heat pump	9600	1/2	—	—	Water pump	335	15	205	4
Heater (radiant)	1300	4	155	3					
Heating pad	60	80	9	0.18					

Cost base is 2 cents per kilowatt-hour (kWH) electricity.

Index